Steffi O. Muhanji • Alison E. Flint • Amro M. Farid

eIoT

The Development of the Energy Internet of Things in Energy Infrastructure

Steffi O. Muhanji
Laboratory for Intelligent Integrated
Networks of Engineering Systems (LIINES)
Thayer School of Engineering,
Dartmouth College
Hanover, NH, USA

Alison E. Flint
Laboratory for Intelligent Integrated
Networks of Engineering Systems (LIINES)
Thayer School of Engineering,
Dartmouth College
Hanover, NH, USA

Amro M. Farid
Laboratory for Intelligent Integrated
Networks of Engineering Systems (LIINES)
Thayer School of Engineering,
Dartmouth College
Hanover, NH, USA

https://doi.org/10.1007/978-3-030-10427-6

To my sisters Ivy and Whitney,
Ivy, you will forever be in my heart. Never to be forgotten. I love you both so much, and I am truly proud of you.

Steffi

Preface

It's been 20 years since Kevin Ashton coined the term the "Internet of Things" (IoT). At the time, the concept was advanced by the Auto-ID Center global research consortium as a means of transforming production and supply chain management. If every product or "thing" could have an RFID tag, then it could potentially "speak" to an RFID reader and provide relevant information like its current location, its production date, and its expected delivery time and location. Products, as they moved through a supply chain, could gain their own sort of "intelligence" through intelligent product agents that negotiated with the rest of the supply chain's entities to reach their final destination. In short, having real-time product-level granularity of an entire supply chain was viewed as a key to a *digitized* industrial revolution called *Industrie 4.0*.

In some ways, a lot has changed. In others, much of this original vision has remained the same. No longer is the Internet of Things solely dependent on RFID tags and readers. Instead, the proliferation of sensor technology in the last two decades has tremendously diversified the notion of IoT to include just about any type of sensor with the potential for connection to a communication network. Similarly, communication networks, particularly wireless ones, have experienced similar leaps in innovation and adoption. For perspective, the Wi-Fi Alliance, the trade association responsible for Wi-Fi technology, was founded in the same year (1999) that the term IoT was first used. Finally, mobile computing devices (like smartphones and tablets) have revolutionized the potential for high computing power near or on edge devices. The associated computing platforms (e.g., Android and iOS) has brought about yet another proliferation of IoT-friendly "apps." This tremendous heterogeneity of new sensors, communication networks, edge computing, and mobile apps has transformed the IoT landscape from its humble beginnings centered on RFID tags and readers. In so doing, IoT has emerged as the dominant new paradigm for the transformation of supply chain operations management.

Why This Book?

However, it would be insufficient to restrict the concept of IoT solely to traditional supply chain management and logistics applications. The Internet of Things now spans every "thing." Among others, there are applications in transportation, water, defense, aerospace, and, yes, even energy systems. This book explores the collision between the sustainable energy transition and the Internet of Things (IoT).

In that regard, this book's arrival is timely. Not only is the Internet of Things for energy applications, herein called the *energy Internet of Things* (eIoT), rapidly developing, but also the transition toward sustainable energy to abate global climate is very much at the forefront of public discourse. The 2016 COP21 Paris Agreement has committed to keep the increase in global average temperature to well below 2 °C. The 2018 report of the Intergovernmental Panel on Climate Change states that achieving such a goal would require "rapid, far reaching, and unprecedented changes in all aspects of society."

It is within the context of these two dynamic thrusts, *digitization* and *global climate change*, that the energy industry sees itself undergoing significant change in how it is operated and managed. This book recognizes that they impose five fundamental energy management change drivers: (1) the growing demand for electricity, (2) the emergence of renewable energy resources, (3) the emergence of electrified transportation, (4) the deregulation of electric power markets, and (5) innovations in smart grid technology. Together, they challenge many of the assumptions upon which the electric grid was first built.

Traditionally, the electricity grid comprised of centralized generation whose soul purpose was to serve consumer demand. This centralized paradigm came to shape the way the electricity grid is managed and operated today. However, as more renewable distributed generations in the form of solar and wind are added to the grid, power can no longer just flow in one direction (from the transmission to the distribution system). Instead, consumers that have rooftop solar should be able to send their power back to the electricity transmission system. Variable renewable energy resources have also put a strain on system operators because they must meet the net load (i.e., consumer demand minus variable energy generation). Furthermore, because many of these variable renewable energy resources are installed behind metering infrastructure, they are not always able to distinguish between the variability of load and that renewable generation. To further complicate the situation, consumers increasingly possess the capability to manage and control their consumption patterns, making it possible for them to respond to the time-of-use or real-time price signals.

Instead of this traditional paradigm of active centralized generation serving passive distributed loads, this book argues that the five energy management change drivers stated above will activate the grid periphery. This will in turn "pull" eIoT technologies to become a scalable energy management solution. In so doing, eIoT will enable a pervasive grid-wide transformation in which a plethora of cyber and physical grid devices will interact within *transactive energy* applications. Energy, power, and other grid "services" will have to be sought in or near real time so as to maintain grid reliability and economic efficiency at all points in a very much distributed grid.

The Goal of This Book

The goal of this book is provide a single integrated picture of how eIoT can come to transform our energy infrastructure. This book links the energy management change drivers mentioned above to the need for a technical energy management solution. It, then, describes how eIoT meets many of the criteria required for such a technical solution. In that regard, the book stresses the ability of eIoT to add sensing, decision-making, and actuation capabilities to millions or perhaps even billions of interacting "smart" devices. With such a large-scale transformation composed of so many independent actions, the book also organizes the discussion into a single multi-layer energy management control loop structure. Consequently, much attention is given to not just network-enabled physical devices but also communication networks, distributed control and decision-making, and finally technical architectures and standards. Having gone into the detail of these many simultaneously developing technologies, the book returns to how these technologies when integrated form new applications for transactive energy. In that regard, it highlights several eIoT-enabled energy management use cases that fundamentally change the relationship between end users, utilities, and grid operators. Consequently, the book discusses some of the emerging applications for utilities, industry, commerce, and residences. The book concludes that these eIoT applications will transform today's grid into one that is much more responsive, dynamic, adaptive, and flexible. It also concludes that this transformation will bring about new challenges and opportunities for the cyber-physical-economic performance of the grid and the business models of its increasingly growing number of participants and stakeholders.

What's in This Book?

This book is comprised of five chapters organized as follows:

- Chapter 1 presents eIoT as a potential solution to the five energy management change drivers described above.
- Chapter 2 recognizes that these drivers will require a transformation of the grid periphery where eIoT is also most suitable as a technical solution.
- Chapter 3 then presents the development of IoT within energy infrastructure using an energy management control loop as a guiding structure for discussion.

- Chapter 4 then ties this overarching techno-economic energy management control loop with the emerging concept of transactive energy. Applications for utilities, industry, commerce, and residences are subsequently discussed.
- Chapter 5 serves to summarize the conclusions of the work. In short,

  1. eIoT will become ubiquitous.
  2. eIoT will enable new automated energy management platforms.
  3. eIoT will enable distributed techno-economic decision-making.

  Chapter 5 also serves to highlight two open challenges and opportunities for future work. These are:

  1. The convergence of cyber, physical, and economic performance
  2. The re-envisioning of the strategic business model for the utility of the future

Hanover, NH, USA Steffi O. Muhanji
Hanover, NH, USA Alison E. Flint
Hanover, NH, USA Amro M. Farid
October 2018

Acknowledgments

The authors would like to thank the Electric Power Research Institute (EPRI) for the partial funding to support this book project. We'd also like to thank EPRI for its technical feedback as this work has developed.

Contents

Nomenclature

Measurement Units

$\$/kWh$	dollars per kilowatt-hour	p. 1
μA	microamps	p. 1
Bps	bits per second	p. 1
BTU	British Thermal Unit	p. 49
$Gbps$	gigabits per second	p. 62
GW	gigawatts	p. 4
Hz	Hertz	p. 78
$kbps$	Kilobits per second	p. 69
kHz	Kilohertz	p. 34
km	kilometers	p. 1
kV	kilovolts	p. 1
kVA	kilo Volt-Ampere	p. 1
kW	kilowatt	p. 24
kWh	Kilowatt-hours	p. 42
m	meters	p. 1
mA	milliamps	p. 1
$Mbps$	megabits per second	p. 62
MHz	Megahertz	p. 65
$MMBTU$	Million British Thermal Units	p. 1
ms	milliseconds	p. 1
$Mtoe$	Million tons of oil equivalent	p. 2
MVA	Mega Volt Amp	p. 87
MW	megawatt	p. 24
ns	nanoseconds	p. 1
pu	per Unit	p. 1
$Quads$	quadrillion BTU	p. 1
s	seconds	p. 1
TWh	Terawatt-hour	p. 112
VAR	Volt-ampere reactive	p. 95
W	watts	p. 1

Acronyms

$3GPP$	Third Generation Partnership Project . . . . . . p. 69
AC	Alternating current . . . . . . p. 41
ACE	Area Control Error . . . . . . p. 35
$ADMM$	Alternate Direction Method of Multipliers . . . . . . p. 78
$ADSL$	Asymmetric digital subscriber line . . . . . . p. 67
AEP	American Electric Power . . . . . . p. 94
AGC	Automatic generation control . . . . . . p. 34
$ALADIN$	Alternating Direct Inexact Newton . . . . . . p. 78
AMI	Advanced Metering Infrastructure . . . . . . p. 43
AMM	Automated meter management . . . . . . p. 43
$AMQP$	Advanced Message Queuing Protocol . . . . . . p. 75
AMR	Automatic meter reading . . . . . . p. 43
$ANSI$	American National Standards Institute . . . . . . p. 41
API	Application Programming Interface . . . . . . p. 75
APP	Auxiliary Problem Principle . . . . . . p. 78
$ARRA$	American Recovery and Reinvestment Act . . . . . . p. 43
ATC	Analytical Target Cascading . . . . . . p. 78
AVR	Automatic voltage regulator . . . . . . p. 34
AWS	Amazon Web Services . . . . . . p. 86
$BB\text{-}PLC$	Broadband power-line communication . . . . . . p. 62
$BEMS$	Building energy-management systems . . . . . . p. 111
BPA	Bonneville Power Administration . . . . . . p. 96
$BPSK$	Binary phase-shift keying . . . . . . p. 66
$BYOT$	Bring Your Own Thermostat . . . . . . p. 112
$CAISO$	California Independent System Operator . . . . . . p. 105
CIM	Common Information Model . . . . . . p. 87
$CoAP$	Constrained Application Protocol . . . . . . p. 75
$CSOC$	Cyber Security Operations Center . . . . . . p. 96
CT	current transformer . . . . . . p. 112
$DACR$	Distribution automation circuit reconfiguration . . . . . . p. 95
DDS	Data Distribution Service . . . . . . p. 75
DER	Distributed Energy Resources . . . . . . p. 9
$DERP$	Distributed Energy Resource Provider . . . . . . p. 105
DG	Distributed Generation . . . . . . p. 9
DMS	Distribution-management system . . . . . . p. 22
$DNP3$	Distributed Network Protocol . . . . . . p. 63
DR	Demand Response . . . . . . p. 9
DSL	Digital subscriber lines . . . . . . p. 67
DSM	Demand Side Management . . . . . . p. 24
DSO	Distribution System Operators . . . . . . p. 104
DVR	Digital video recorder . . . . . . p. 46
EIA	Energy Information Administration . . . . . . p. 45
$eIoT$	Energy Internet of Things . . . . . . p. 1
$EISA$	Energy Independence Security Act . . . . . . p. 83

EMS	Energy-management system . . . . . p. 22
EPRI	Electric Power and Research Institute . . . . . p. 32
ESCO	Energy service company . . . . . p. 71
ETSI	European Telecommunications Standards Institute . . . . p. 65
EV	Electric vehicle . . . . . p. 4
FACTS	Flexible AC Transmission System . . . . . p. 34
FAN	Field Area Network . . . . . p. 66
FCC	Federal Communications Commission . . . . . p. 66
FERC	Federal Energy Regulatory Commission . . . . . p. 35
FIT	Feed-In Tariff . . . . . p. 4
FRA	frequency response analysis . . . . . p. 2
G3-PLC	3rd Generation Power-Line Communication . . . . . p. 62
GHG	Greenhouse gas . . . . . p. 5
GPRS	General Packet Radio Service . . . . . p. 69
GPS	Global Positioning System . . . . . p. 34
GSM	Global System for Mobile . . . . . p. 65
GWAC	GridWise Architecture Council . . . . . p. 84
HAN	Home Area Network . . . . . p. 66
HART	Highway Addressable Remote Transducer . . . . . p. 73
HDSL	High-bit-rate digital subscriber line . . . . . p. 67
HEM	Home energy management . . . . . p. 65
HTTP	HyperText Transfer Protocol . . . . . p. 75
HVAC	Heating Ventilation and Air-Conditioning . . . . . p. 46
ICT	Information and Communication Technologies . . . . . p. 87
IEA	International Energy Agency . . . . . p. 2
IEC	International Electrotechnical Commission . . . . . p. 80
IEEE	Institute of Electrical and Electronics Engineers . . . . . p. 86
IEEE-PES-DMS	IEEE Power and Energy Society Distribution Automation/Management System . . . . . p. 20
IETF	Internet Engineering Task Force . . . . . p. 75
IFTTT	If This Then That . . . . . p. 112
IIoT	Industrial Internet of Things . . . . . p. 73
IoT	Internet of Things . . . . . p. 1
IP	Internet Protocol . . . . . p. 64
IPP	Independent Power Producers . . . . . p. 19
IPSO	Internet Protocol for Smart Objects . . . . . p. 11
IPv6	Internet Protocol version 6 . . . . . p. 11
ISO	Independent System Operator . . . . . p. 9
IT	Information Technology . . . . . p. 43
ITC	Investment Tax Credit . . . . . p. 5
ITES	Intelligent Transportation Energy System . . . . . p. 53
ITU	International Telecommunication Union . . . . . p. 62
ITU-T	ITU Telecommunications Standardization Sector . . . . . p. 62
KAIST	Korea Advanced Institute of Science and Technology . p. 11
KKT	Karush–Kuhn–Tucker . . . . . p. 78

LAN	Local Area Network	p. 54
LMP	Locational Marginal Price	p. 101
LoRa	Long Range	p. 65
LoRaWAN	Long Range Wide-Area Network	p. 65
LPWAN	Low-power wide-area network	p. 65
LTC	load tap changer	p. 2
LTE	Long-Term Evolution	p. 56
M2M	Machine-to-machine	p. 20
MAC	Medium Access Control	p. 65
MAS	Multi-agent system	p. 79
MASCEM	Multi-agent system competitive electricity markets simulator	p. 80
MATSIM	Multi-Agent Transport Simulation	p. 52
MIS	metal-insulated semiconductor	p. 2
MIT	Massachusetts Institute of Technology	p. 11
MQTT	Message Queue Telemetry Transport	p. 75
MTU	Master Terminal Unit	p. 31
N2N	NAN-to-NAN	p. 70
NAN	Neighborhood Area Network	p. 54
NB-IoT	Narrow band Internet of Things	p. 65
NB-PLC	Narrowband power-line communication	p. 62
NIST	National Institute of Standards and Technology	p. 88
NOPR	Notice of Proposed Rule-making	p. 105
NWS	Non-wires Solutions	p. 94
OASIS	Organization for Advancement of Structured Information Standards	p. 93
OECD	Organization for Economic Cooperation and Development	p. 2
OPF	Optimal Power Flow	p. 78
OPP	Olympic Peninsula Project	p. 94
P2P	Peer-to-peer	p. 101
PDC	Phasor data concentrator	p. 64
PHEV	Plug-in electric vehicles	p. 4
PHY	Physical layer	p. 65
PLC	Power-line carrier	p. 62
PLC	Programmable logic controllers	p. 31
PMU	Phasor Measurement Unit	p. 34
PNNL	Pacific Northwest National Laboratory	p. 94
PNWSGD	Pacific Northwest Smart Grid Demonstration	p. 94
PRIME	PoweRline Intelligent Metering Evolution	p. 62
PROFIBUS	Process Field Bus	p. 63
PT	potential transformer	p. 2
PTC	Production Tax Credit	p. 5
PV	Photovoltaic	p. 3
QB-PLC	Quasi-band power-line communication	p. 62

QoS	Quality of service . . . . . p. 70
RE	Renewable Energy . . . . . p. 113
RF	RF . . . . . p. 2
$RFID$	Radio-frequency identification . . . . . p. 12
ROI	Return-on-Investment . . . . . p. 47
$RPMA$	Random Phase Multiple Access . . . . . p. 66
RPS	Renewable energy portfolio standards . . . . . p. 4
RTO	Regional Transmission Organization . . . . . p. 105
$RTPda$	Real time pricing with double auction . . . . . p. 96
RTU	Remote Terminal Units . . . . . p. 31
$SCADA$	Supervisory control and data acquisition . . . . . p. 20
$SCED$	Security-constrained economic dispatch . . . . . p. 39
$SCUC$	Security-constrained unit commitment . . . . . p. 39
SDK	Software Developer's Kit . . . . . p. 86
SDR	Software-defined radio . . . . . p. 75
$SEDC$	Smart Energy Demand Coalition . . . . . p. 105
$SGAM$	Smart Grid Architecture Model . . . . . p. 83
$SGIG$	Smart Grid Investment Grant . . . . . p. 43
$SGIR$	Smart grid interoperability reference model . . . . . p. 87
SIA	Seamless Integration Architecture . . . . . p. 87
$SIWG$	Smart Inverter Working Group . . . . . p. 41
SND	Software-defined networking . . . . . p. 75
SSL	Secure Socket Layers . . . . . p. 75
$STATCOM$	Static Synchronous Compensator . . . . . p. 37
SVC	Static VAR Compensator . . . . . p. 37
TCP	Transmission Control Protocol . . . . . p. 75
$TCPST$	Thyristor Controlled Phase Shifting Transformer . . . . . p. 37
$TCSC$	Thyristor Controlled Series Compensator . . . . . p. 37
TE	Transactive Energy . . . . . p. 9
$TeMIX$	Transactive Energy Market Information Exchange . . . . . p. 93
TEN	Transportation Electricity Nexus . . . . . p. 52
TLS	Transport Layer Security . . . . . p. 75
TSO	Transmission system operators . . . . . p. 108
TSP	Transactive Systems Program . . . . . p. 93
UHF	Ultra-high frequency . . . . . p. 63
UL	Underwriters Laboratory . . . . . p. 41
$UNB\text{-}PLC$	Ultra-narrowband power-line carrier . . . . . p. 62
$USDOE$	United States Department of Energy . . . . . p. 43
UTP	Unshielded Twister Pair . . . . . p. 71
$VDSL$	Very-high-bit-rate digital subscriber line . . . . . p. 67
VER	Variable energy resources . . . . . p. 38
$VVAR$	Volt VAR Optimization . . . . . p. 95
$WAMS$	Wide-Area Monitoring Systems . . . . . p. 34
WAN	Wide-Area Network . . . . . p. 54
$WiMAX$	Worldwide Interoperability for Microwave Access . . . . . p. 69

$WiSUN$	Wireless Smart Utility Network . . . . p. 66
WPP	Wind power plants . . . . p. 40
WSN	Wireless sensor networks . . . . p. 24
XML	Extensible Markup Language . . . . p. 75
$XMPP$	eXtensible Messaging and Presence Protocol . . . . p. 75

List of Figures

List of Tables

Executive Summary

The electric power grid was developed on an architectural assumption of centralized generation being delivered to passive distributed loads irrespective of the cost required to do so [33]. However, several new energy-management change drivers are emerging to uproot this status quo. Chapter 1 identifies these drivers as the rising demand for electricity [34–36], the emergence of renewable energy resources [37–40], the emergence of electrified transportation [41, 42], the deregulation of power markets [43, 44], and innovations in smart grid technology [45, 46]. Responding to these drivers requires new and integrated technical solutions for energy management.

The energy Internet of Things (eIoT) has been proposed as one such energy-management solution, illustrated in Fig. 1. eIoT is a leading and overarching perspective where all devices that consume electricity are internet-enabled and, consequently, can coordinate their energy consumption with the rest of the grid in or near real-time. eIoT technologies must, therefore, be adopted within the context of these energy-management change drivers.

Perhaps nowhere will the impact of the energy-management change drivers identified in the previous paragraphs be felt more than at the grid's periphery. Distributed generation (DG), in the form of solar photovoltaics (PV) and small-scale

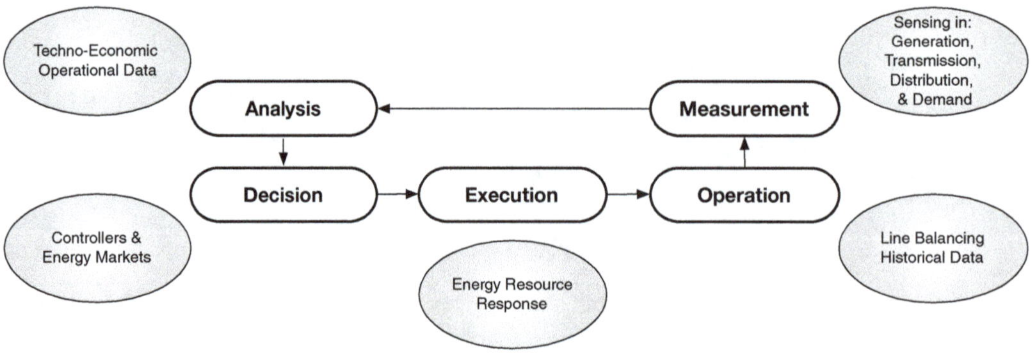

Fig. 1 A closed-loop framework for electrical power system management

wind, will be joined by a plethora of internet-enabled appliances and devices to transform the grid's periphery to one with two-way flows of power and information [45, 46]. This transformation is a daunting technical challenge. Not only are there tens of millions of devices at the leaves of the grid's radial structure, these devices are relatively small and require innovations in sensing, communication, control, and actuation. Chapter 1 first describes this transformation and then describes the challenge of activating the grid's periphery. Finally, it describes how eIoT can potentially be deployed as a scalable energy-management solution.

The development of IoT within energy infrastructure is best seen as a control loop. The control loop is composed of four functions: a physical process (such as the generation, transmission, or consumption of electricity), its measurement, decision-making, and actuation. This control structure is shown in Fig. 2 where a sensor takes measurements of the physical system's states and outputs. Wireless and wired communications are then used to pass this information between the physical layer and other informatic components. This information is used to make decisions either independently in a decentralized fashion or in coordination with the informatic components of other devices. Decisions are then sent back down to network-enabled actuators for implementation. In some cases, this control loop acts in near real-time. In other cases, some of the information is used as part of predictive applications that facilitate decisions at a longer time scale. Control algorithms implemented at different layers of this control loop enable the control of individual devices as well as the coordination of smart grid devices that comprise other parts of eIoT. Given the connectivity between the functions of this control loop, its successful implementation requires architectures and standards that ensure interoperability between eIoT technologies.

Chapter 3 serves to summarize the most recent developments of IoT within energy infrastructure. The discussion proceeds from the bottom-up by classifying these developments according to the generic control structure shown in Fig. 2.

- Section 3.1 discusses some of the state of the art in network-enabled physical devices, whether they are network-enabled sensors or actuators in the control loop. Section 3.2 then focuses on the communication networks that send and receive data to and from these devices.
- Section 3.3 then discusses advancements in distributed control algorithms that coordinate the techno-economic performance. The chapter concludes with two discussions of a cross-cutting nature.
- Section 3.4 addresses the importance of control architectures and standards in the development of eIoT technologies.
- Section 3.5 addresses the security and privacy concerns that emerge from the development of eIoT technologies.

When these many factors are implemented together properly, they form an eIoT control loop that effectively manages the technical and economic performance of the grid. This control loop is most consonant with the emerging concept of "transactive energy" (TE), which is commonly viewed as a collection of techniques to manage the exchange of energy in business transactions [47]. A utility, or any other private jurisdiction, can implement TE between its various customers in industrial, commercial and residential environments to manage distributed energy resources (DERs) technologies. TE applications incorporate the new eIoT-based activities for utilities and for industrial, commercial, and residential consumers. The result is better management of resources, successful integration of renewable energy, and increased efficiency in grid operations [47]. In many ways, TE is seen as an effective way to manage the technical and economic performance of various grid operations at all levels of control—commercial, industrial, or residential. As such, eIoT technologies directly support the implementation of TE applications.

Chapter 4 discusses how aspects of the eIoT control loop from Chap. 3 are reflected in various TE applications across different layers of the electricity value chain:

- Section 4.1 discusses the role of TE in future grid applications and highlights some of the proposed TE frameworks.
- Section 4.2 presents a few motivational use cases for TE frameworks.
- Section 4.3 then addresses the role of the utility and distribution system operators (DSOs) within the TE framework. This section also recognizes some of the challenges and opportunities presented by the implementation of TE.
- Finally, Section 4.4 examines various customer applications for TE and eIoT in commercial, industrial, and residential settings.

In conclusion, the development of eIoT is an integral part of the transformation to the future electricity grid. It will transform all aspects of grid operations and control. This transformation spans both technical and economic layers and leads to

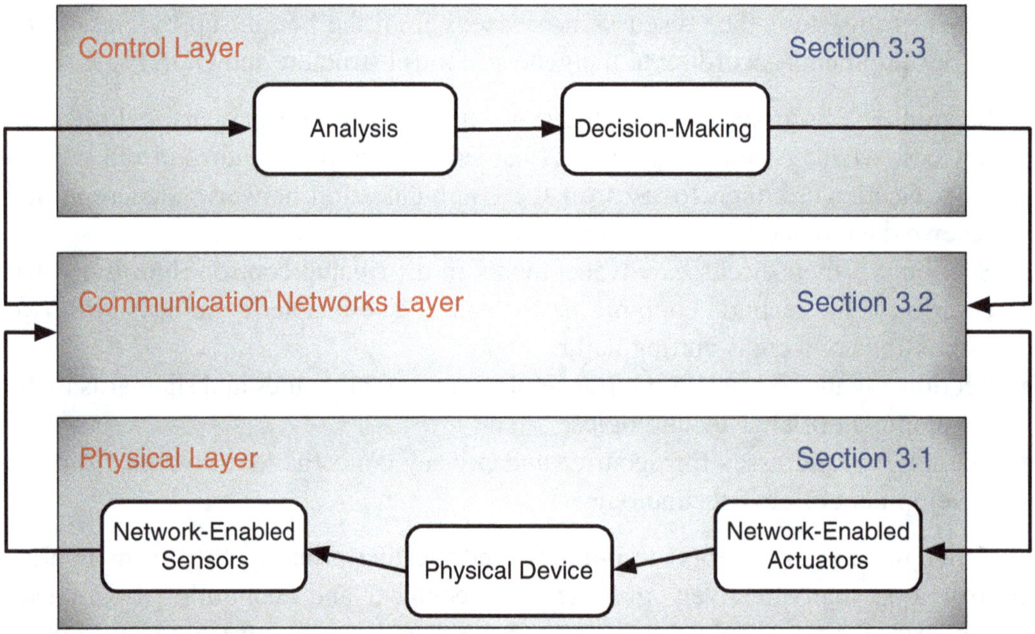

Fig. 2 The development of IoT within energy infrastructure as networked control loop

new applications, stakeholders, and energy system management solutions. Chapter 5 serves to summarize the conclusions of the work. In short,

1. eIoT will become ubiquitous.
2. eIoT will enable new automated energy management platforms.
3. eIoT will enable distributed techno-economic decision-making.

Chapter 5 also serves to highlight two open challenges and opportunities for future work. These are:

1. The convergence of cyber, physical, and economic performance
2. The re-envisioning of the strategic business model for the utility of the future

Chapter 1
eIoT as a Solution to Energy-Management Change Drivers

The electric power grid was developed on the architectural assumption of centralized generation being delivered to passive distributed loads irrespective of the cost implication [33]. However, several new energy-management change drivers have emerged to uproot this status quo. These drivers include a rising demand for electricity [34–36], the emergence of renewable energy resources [37–40], the emergence of electrified transportation [41, 42], deregulation of power markets [43, 44], and innovations in smart grid technology [45, 46]. Responding to these drivers requires new and integrated technical solutions for energy management.

The internet of things (IoT) for energy applications, herein called the "energy internet of things" (eIoT), has been proposed as one such energy-management solution, illustrated in Fig. 1.1. eIoT is a leading and overarching perspective where all devices that consume electricity are internet-enabled and consequently can coordinate their energy consumption with the rest of the grid in real time or near real time. eIoT technologies must, therefore, be adopted within the context of these emerging energy-management change drivers.

1.1 Energy-Management Change Drivers

Several change drivers are causing a fundamental shift in energy-management practices in the electric power grid. These change drivers include:

- Growing demand for electricity,
- Emergence of renewable energy resources,
- Emergence of electrified transportation,
- Deregulation of electric power markets,
- Innovations in smart grid technology.

S. O. Muhanji et al., *eIoT*, https://doi.org/10.1007/978-3-030-10427-6_1

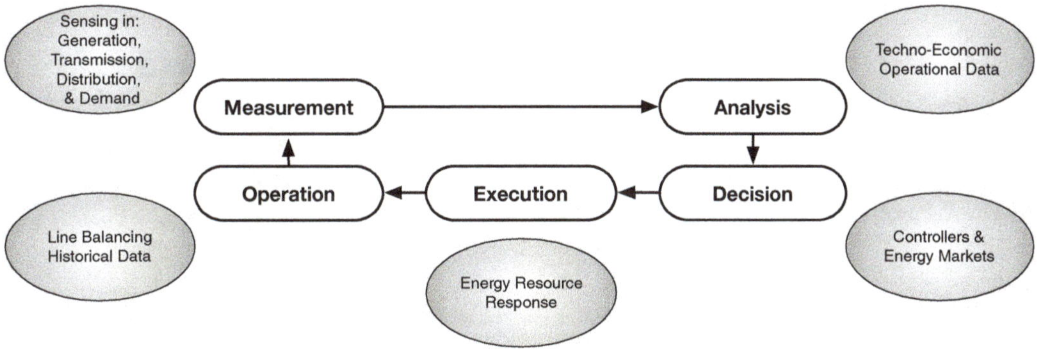

Fig. 1.1 A closed-loop framework for electrical power system management

1.1.1 *Growing Demand for Electricity*

The first of these drivers is the rising global demand for electricity which follows a larger global trend where the demand for all types of energy in developing countries is growing. The International Energy Agency's (IEA) 2016 World Energy Outlook Report projects the growth of Total Primary Energy Demand from 1161 million tons of oil equivalent (Mtoe) in 2014 to between 1705–2017 Mtoe in 2025 and 2528–4049 Mtoe in 2040 [48]. During that time, global electricity consumption is projected to increase by around 2% per year [48]. Demand for electricity in industrializing economies outpaces renewable electricity generation so that displacement does not occur, but energy generation from all available sources continues to grow [48].

Meanwhile, in developed countries, electricity demand will continue to grow. Although in recent years electricity demand has been nearly flat in many developed countries, electric load growth is expected to return in order to support fuel-switching and other decarbonization trends [49, 50]. Figure 1.2 shows that most of the energy growth will occur in developing countries that are outside the Organization for Economic Cooperation and Development (OECD) countries. Furthermore, during that time, renewable generation growth will increase more quickly than demand and is expected to replace fossil-fuel generation [48]. As a result, any advancement made to accommodate renewable energy in countries with existing infrastructure will have a profound impact on the world's decarbonization efforts.

1.1.2 *The Emergence of Renewable Energy Resources*

The growth and widespread adoption of renewable energy resources is expected to significantly alter the generation mix. This widespread adoption is encouraged by advanced research, state-of-the-art technologies, and favorable legislation that

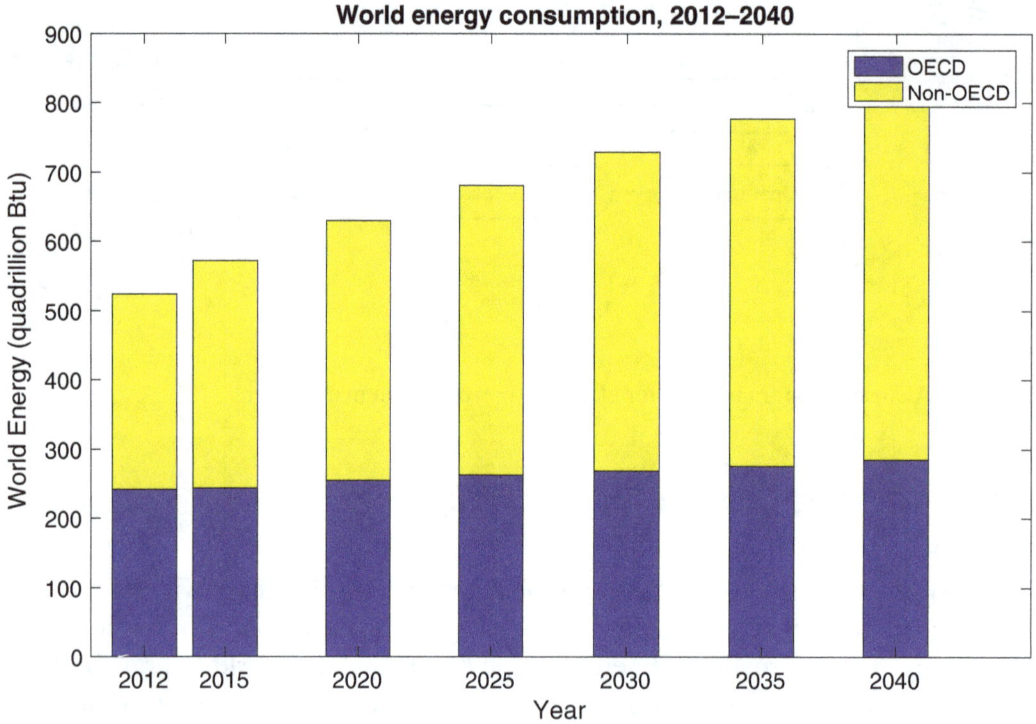

Fig. 1.2 World energy growth between 2015 and 2040 [1]

continue to improve renewable energy resources. These factors have advanced wind and solar technologies, and have pushed them to become more efficient and cost-effective as compared to thermal generation. Research in new wind turbine designs has resulted in improved turbine efficiency and wind power output [51–53]. With these improvements, the cost of wind generation is set to decrease significantly. In fact, the IEA projects that the average costs for wind generation will decline by 15% for onshore wind and by one third for offshore wind between 2017 and 2022 [54].

Further research in solar cell technologies has also led to much higher conversion efficiencies for solar cells. For example, the efficiencies of commercial mono- and poly-crystalline solar modules increased from 12–14% in 2006 to 16–18% in 2016, while that of high-efficiency N-type modules reached an efficiency of over 21% [54]. In addition, generation costs for utility photovoltaic (PV) solar are expected to fall by one-quarter over the period 2017–2022 [54, 55].

Similarly, the growing amount of new legislation and regulations favoring generation and supply of clean energy has forced the evolution of the electricity supply infrastructure and operations to support renewable energy sources. Favorable policies have not only helped lower the cost of investment in these technologies but they have also created competitive market environments for solar and wind projects [54]. Two developed countries and the European Union (EU), in particular, display how renewable energy policy is setting a precedent for countries where the energy infrastructure has yet to reach maturation. Favorable legislation in China and the United States (USA) has played a key role in promoting the widespread adoption of renewable energy resources [56]. These legislations and a commitment towards decarbonization have encouraged investments in renewable energy resources for both small-scale consumers and large-scale energy developers.

Legislation initiatives in China have made a strong impact on the growth of the country's renewable energy capacity [57]. China is projected to add up to 1300 gigawatts (GW) of generation by 2040, which more than doubles its combined growth of fossil fuel and nuclear power capacity [48]. In part to cut back air pollution, China has set 5-year plans to reach 2020 renewable energy targets [56]. As of 2017, China had surpassed its solar PV target and is estimated to meet its wind target by 2020 [54, 56]. These targets have helped China achieve over 40% of global renewable capacity growth by 2016 [54]. By the end of 2015, China's cumulative installed wind capacity was 180.4 GW with 30.5 GW alone being installed in 2015 [58]. Despite these installations, China still faces many challenges towards the growth of renewable energy resources such as the uneven distribution of capacity and unmatched economic growth [58]. China remains the world's largest solar cell producer and consumer [59], a position it has held since 2009. As of December of 2015, China's installed PV capacity was 43.18 GW accounting for 14.9% of the global solar PV capacity [58]. Solar PV installations are expected to continue growing with one study predicting the total installed capacity of 200 GW by 2030 [58].

Developments in wind and solar in China are supported by either a national feed-in tariff (FIT) program or direct subsidies that are meant to encourage the deployment of these resources [58, 59]. Overall, China's central government has guided participation by developers and financial stakeholders to foster large-scale investment in renewable energy [60]. Soon, due to an increase in energy subsidies and integration costs, China is expected to adjust its policies to a quota system with green certificates [54]. Going forward, however, it is still unclear how this shift in legislation will affect the country's overall renewable energy growth and decarbonization efforts. That said, there are still many challenges facing the growth of renewable energy resources, such as uneven distribution of capacity and unmatched economic growth. For example, inner Mongolia has 28% of the over installed wind capacity despite having a low demand of just 6.78% [58]. While areas like Zhejiang, Fujian, and Guangdong province that have a higher population density and contribute 20.5% of the consumer load only have 4.7% of the installed capacity [58]. These disparities in capacity distribution present operational challenges that may influence future renewable legislation in China.

The USA experienced fast growth in wind and solar technologies primarily due to: (1) renewable energy portfolio standards (RPS), (2) state-level policies supporting distributed solar PV and electric vehicles (EVs), and (3) federal tax credits for wind and solar industries [54]. As of 2015, the tax credit for wind producers was 2.3 cents per kilowatt-hour, and solar power developers still receive tax credits for 30% of the value of their investment [61]. Both tax credits are set to expire in 2020, but a 2016 tax bill proposition began phasing out wind credits starting in 2017 [62, 63] and completely terminated solar credits. As per the new tax bill, on the production tax credit (PTC) is gradually phased down for wind and is expired for other technologies such as solar, biomass, and geothermal, for projects beginning construction after December 2016. The PTC will be subject to a 20% step-down in 2017, 40% in 2018, and 60% in 2019 [63, 64]. A similar phase-out schedule applies to the wind energy investment tax credit (ITC), where the allowable tax credit is 30% of expenditures in 2016, 24% in 2017, 18% in 2018, and 12% in 2019 [63, 65]. Although the future of federal tax credits is uncertain, the USA is the second-largest growth market for renewable energy generation sources after China [54].

Most of these changes are happening at the state level with states such as California and New York taking a lead on decarbonization efforts. For several states, the goal is to reach 40% decarbonization (50% for California) by 2030 and 80% by 2050 [66–68]. Decarbonization efforts have focused largely on increasing the renewable energy capacity and energy efficiency improvements, but, lately, these efforts are shifting to include electrified transportation and electric indoor heating [67, 68]. Recently, new regulation by the Federal Energy Regulatory Commission (FERC) has allowed the participation of distributed energy resources in electricity wholesale markets [69]. This regulation will not only improve the deployment of DERs but will also enable the creation of market structures that are more inclusive for DERs.

In the EU, there is a strong interest in wind energy. However, investment has lagged behind due to the lack of support for investments by non-member states [70]. Progress in the deployment of wind technologies is contingent upon the creation of a favorable policy framework that helps bridge this gap in investment [70]. In 2009, the 2009/28/EC Directive to promote the use of renewable energy was adopted by the European Parliament and the Council of Ministers. The directive promoted the development of renewable energy sources as one of the main objectives of the EU energy policy [71]. It also set mandatory national targets that would ensure at least a 20% renewable energy share in total energy consumption by 2020 [70, 71]. By June 2010, each member state was required to have a national plan that defined the technology mix scenario, the trajectory to be followed, and the measures and reforms necessary to overcome barriers and to enable the development of renewable energy [70]. Wind energy was a main component in these national energy plans with an estimated 209.6 GW of wind capacity to be installed by 2020 within the EU [70]. This accounted for 43.1% of the expected renewable energy technologies installed by 2020 [70]. Nevertheless, the EU remains on track to meet their goal of reaching 20% renewable generation by 2020 [72].

A recent report by the renewable energy agency shows that the EU has been able to cut its associated greenhouse gas (GHG) emissions by fossil-fuel generation by about one-tenth [72]. The share of the renewable energy in the total energy consumed in the EU was reported to be 17% in 2016 from the 16.7% reported in 2015 [72]. These numbers show that the EU is likely to still meet its 2020 decarbonization target. However, the stability of the policy framework still remains a potential barrier to meeting this goal for wind energy investors [70]. In future frameworks, policies must address cooperation among nations within and outside of the EU membership [71]. Furthermore, cooperation between countries in renewable energy development projects is imperative for the EU in terms of technical exchanges, economic ties, and political relationships [71].

1.1.3 The Emergence of Electrified Transportation

Third, the new load from electric vehicles requires fundamental upgrades to the electricity infrastructure. New advancements in EV batteries and fast charging technology have led to reduced costs of electric vehicles. A recent review puts the costs per kWh of an electric vehicle battery pack at $500 [73]. This cost is estimated to be even lower (≈$300) for vehicle manufacturers [73]. Although this cost needs to fall to below $150/kWh for electric vehicles to be as price competitive as gasoline vehicles, these lower costs have made electric vehicles much more accessible and affordable [73].

In addition to improved technologies, many countries have adopted electric vehicle mandates to promote EVs and reduce the CO_2 emissions of their transportation system. Countries including China, the UK, France, India, and Norway have national legislation to encourage the sale and production of EVs [74]. As a result, car makers are responding with large monetary investments into electrifying their fleets [75]. Although many countries will not establish similar policies, these large mandates are set to contribute to a competitive environment for EVs internationally. Consequently, the falling costs of vehicles will affect the US consumers and encourage the integration of EV infrastructure into the US electricity grid.

In the USA, federal income tax credits and state-level cash incentives are available to consumers who purchase electric vehicles [76]. For example, a federal income tax credit of $7500 is available for vehicles delivered before the end of 2018 and over 13 states offer cash incentives to consumers [76]. In addition to cash incentives, other non-cash incentives such as carpool lanes and free municipal parking are offered by some states to EV owners [76]. These incentives have largely contributed towards the widespread adoption of EVs.

The future fleet of EVs requires a large load of energy that the current electricity system does not produce or support. Most EVs require around 0.2–0.3 kWh of charging power per mile of driving [3]. A plug-in vehicle of 1.4 kW more than doubles the average evening load of a household, and fast chargers, at 6.6 kW or higher, will significantly alter the load pattern of the consumer [3].

On an energy basis, the electrification of transport will have a substantial impact on the current capacity of the electric power grid. One study estimates that with a 100% electrification of transport by 2050, the total electricity demand will increase by 2100 TWh [77]. This represents 56% of the 2015 electricity sales [77]. Consider Fig. 1.3. In 2016, the USA consumed 27.9 quads (quadrillions, or Btu $\times 10^{15}$) of energy whereas the electric power grid only delivered 12.6 quads of useful electricity. Such a figure suggests that the electric power grid will require significant upgrades in order to accommodate a large-scale electrification of transportation. Furthermore, electrified transportation has the potential to complicate power system operations—in balancing, line congestion, or voltage control [78, 79].

Figure 1.4 shows the potential impact of plug-in electric vehicles on residential customers' electrical load. Beyond the need for higher rated electrical panels in the home, several plug-in vehicles could overload distribution circuits and transformers

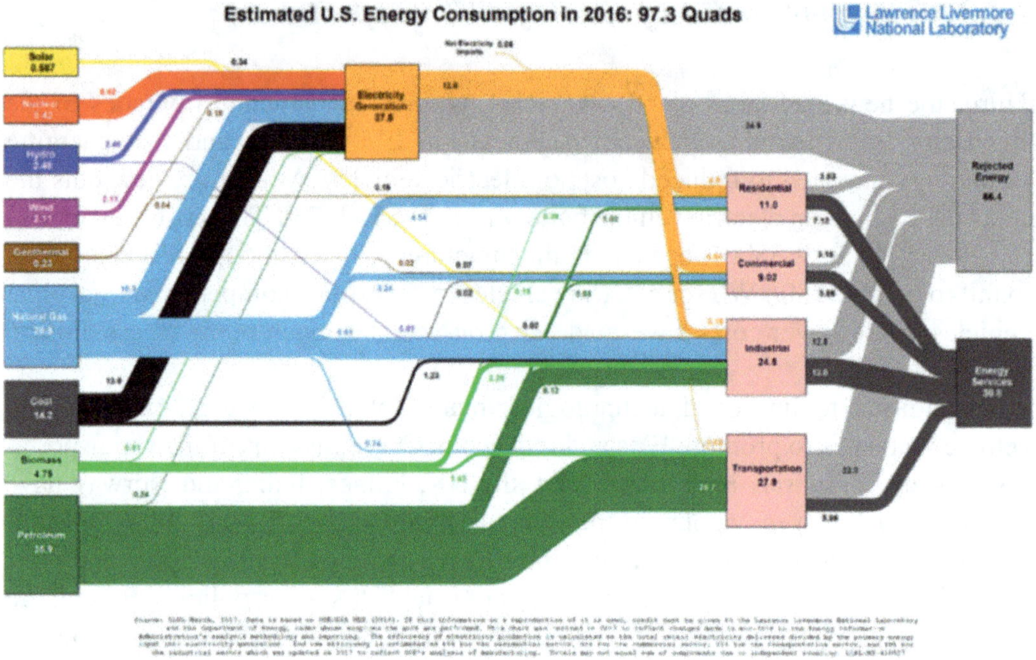

Fig. 1.3 Sankey diagram of American energy system in 2016 [2]

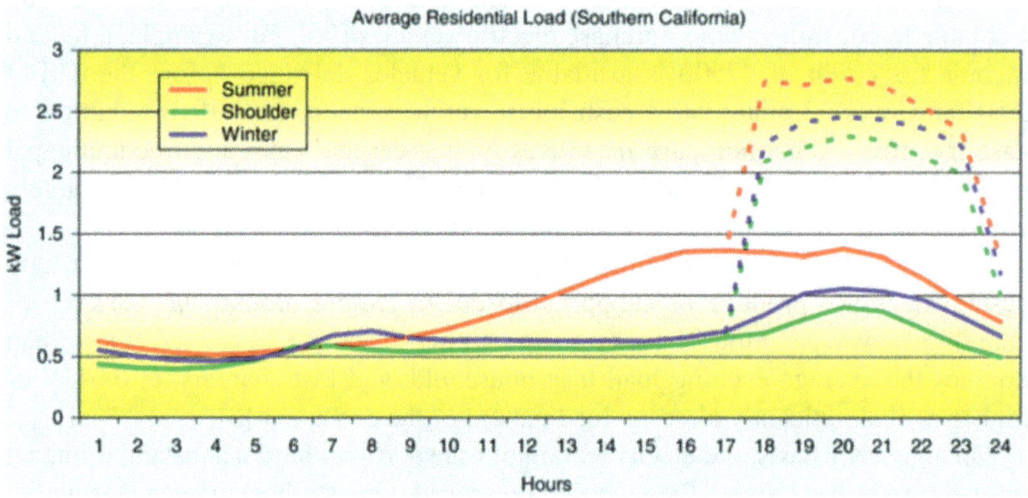

Fig. 1.4 Plug-in EVs as a new and significant component of residential consumer load [3]

that normally operate close to their limits [3]. With normal demand variations, several plug-in vehicles may overload a 25- or 50-kVA secondary transformer on a single-phase lateral [3]. EV loads can also create unbalanced conditions on distribution system feeders [3]. Therefore, advanced control strategies for charging EVs such as coordinated charging [80, 81], vehicle-to-grid stabilization [79, 82–86], and charging queue management [87, 88] have been proposed to stabilize electric vehicles' charging schedules. These works have determined that a holistic approach to studying electric vehicles is necessary given the coupling with the electricity sector [31, 89–91]. Electrified transportation is discussed further in Sect. 3.1.5.7.

1.1.4 Deregulation of Electric Power Markets

Fourth, during the deregulation trend of the 1990s, American power markets were restructured so as to become more diversified and competitive [44, 92–95]. Figure 1.5 shows a transition from a fully regulated (monopolistic) electric power system to one that is fully deregulated [96]. Debundling generation, transmission, and distribution was intended to lower customer rates and improve the quality of service [44]. Utility activities in resource production have also become deregulated, thus opening resource trading on wholesale markets by non-traditional parties [97]. Presently, energy retailers interact directly with customers, and in countries with high regulation, the distribution network operator takes on the role of a service aggregator [97].

More recently, there has been steady progress towards the development of deregulated markets in the distribution system as well [98, 99]. Data services present in physical transmission and distribution are typically unregulated, and IoT can facilitate supply-chain management as well as demand-side market participation [97]. As a result, companies that offer aggregation services may play a larger role in selling distributed power at both the local and wholesale level.

Continuing on the trend towards deregulation, transactive energy (TE) has been proposed as a means of managing generation and demand through the use of time-dependent economic constructs while giving adequate consideration to reliability [100]. In many ways, it is considered a new "smart grid" approach to synthesize measurements, devices, and market information into an emerging fair market for the electricity grid [101]. This market requires real-time data, interconnection among systems, and judicial transparency of information and market operations [101].

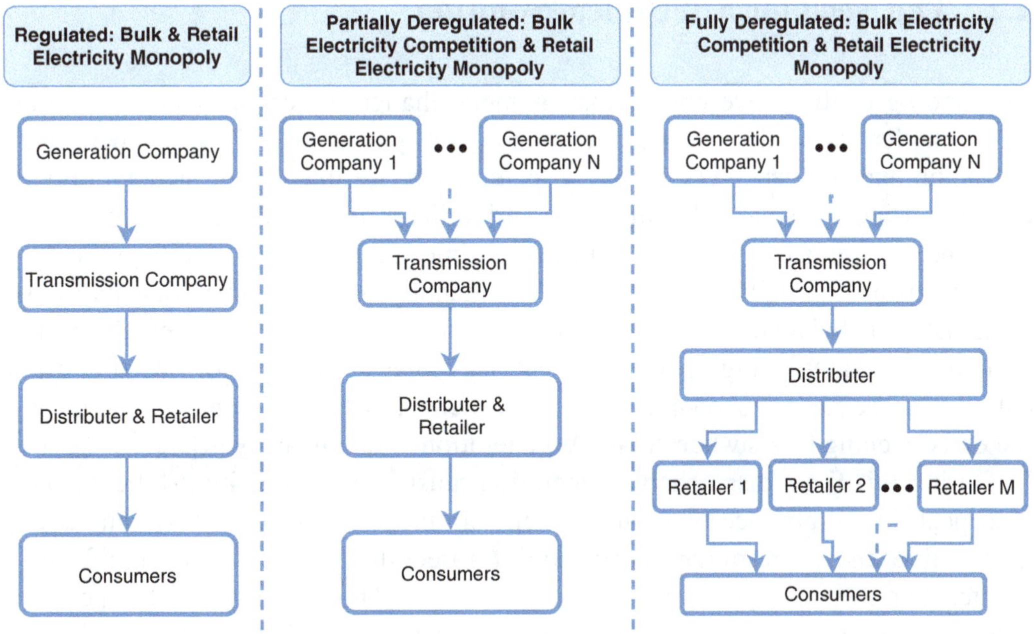

Fig. 1.5 Types of regulated and deregulated environments

TE approaches can establish distributed energy resources (DERs) in energy markets, and further liberate consumer choice in power services. However, techniques for measurement, market surveillance, and market contract enforcement are necessary for expanding the number of market participants [101], which easily exemplifies how market complexity can increase rapidly. TE, which is discussed at length in Chap. 4, is perhaps one of the most compelling use cases for eIoT.

1.1.5 Innovations in Smart Grid Technology

In recent years, the electric power system has seen a steady stream of new "smart" technology innovations [102–104]. Although these innovations enable new functions and services, they also increase the operational complexity of the grid [105–107]. A *smart grid* is commonly defined as a power system that allows two-way communication and two-way flow of power [106] through advanced control and decision-making functionality. It supports decentralized energy generation where power is injected from the grid periphery back into the larger electrical power system. This brings about many opportunities in distributed generation (DG), distributed energy resources (DER), demand response (DR) as well as TE. These technological innovations are quickly transforming the structure and function of the electric power grid. Consequently, pricing mechanisms and regulatory bodies must keep pace with this rapid technological transformation by creating appropriate framework adjustments and legislation to standardize the grid's development [46, 106].

1.2 The Need for a Technical Solution

Responding to these five energy-management change drivers presents new reliability challenges to the overall operation of the power grid. In grid operations, balancing and frequency control are affected by renewable energy generation (for example, wind and solar PV). Due to the variability of renewable energy generation, grid operators must now dispatch to a real-time load profile that is significantly different from the daily load profile. Consequently, the grid operators may have to adjust their balancing operations to accommodate this new requirement on the system. For example, high penetration rates of solar PV bring about what is often called a "duck curve" (shown in Fig. 1.6), which exhibits a very sharp ramp during the early evening hours when solar PV generation is fading away [4].

During this time, dispatchable generation must respond quickly to the evening load peak in the absence of solar PV generation. Solar PV and wind generation, as variable energy resources, also exhibit forecast errors that are significantly greater than the forecast errors for load [108, 109]. This is partly due to operators having many more decades of experience forecasting load than wind and solar PV generation. The larger forecast errors further complicate balancing operations.

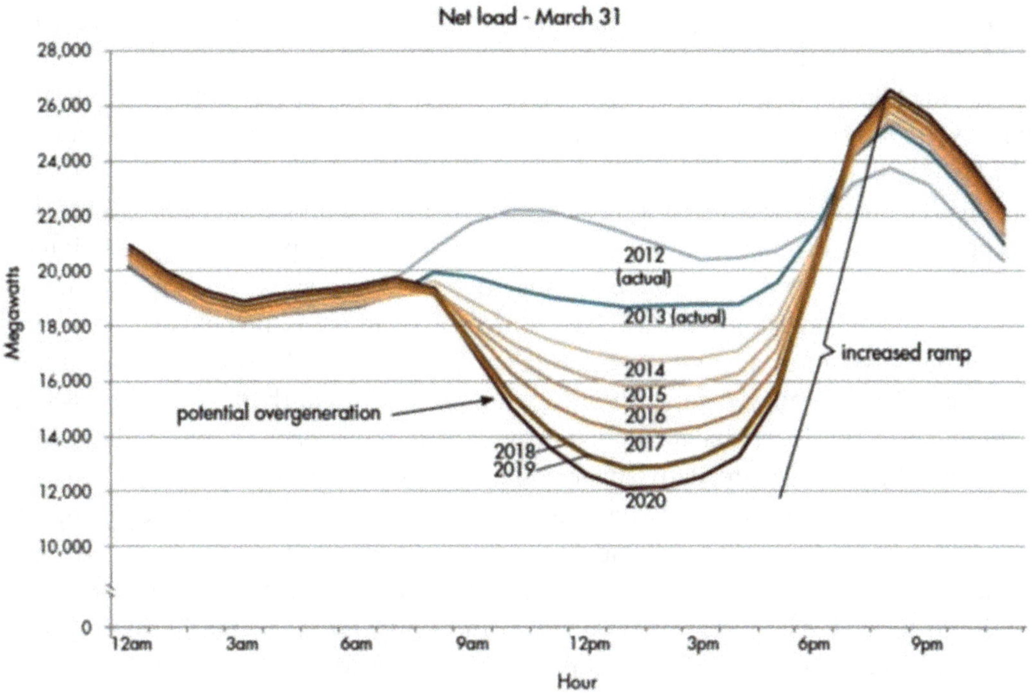

Fig. 1.6 The California ISO duck curve [4]

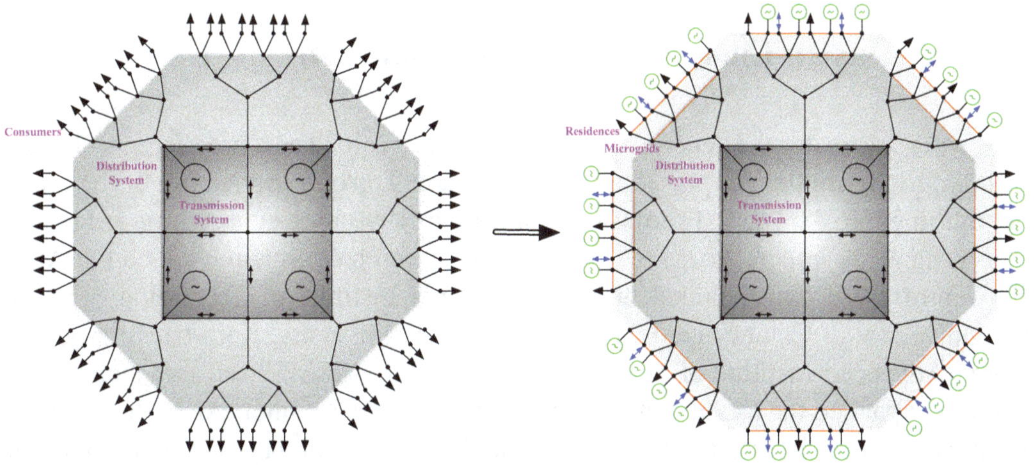

Fig. 1.7 A conceptual transition from a traditional electric power grid to a future smart grid

In addition to these challenges in balancing operations, much renewable energy is integrated as distributed generation at the periphery of the electric power system (see Fig. 1.7). Currently, the electric distribution system is designed for one-way flow of power out to consumers [33, 110]. The presence of distributed generation creates the potential for two-way power flow in the distribution system. Consequently, the distribution system's protection equipment must be redesigned to accommodate two-way flow of power [111].

Furthermore, the widespread integration of DG on a radial topology has the potential to exceed transformer ratings [112, 113] and/or exceed line flow limits in this backward direction. Hence, when adding two-way power flow from variable energy resources, voltage limits, phase balances, and load balancing are threatened [114].

Finally, the distribution system was designed for a monotonically decreasing voltage profile from generation down to the load. The presence of distributed generation at the grid periphery can cause over-voltages as power flows upstream towards the transmission system. These structural changes to the physical grid bring about new dynamics at multiple timescales. Within seconds to minutes, ancillary services like frequency regulation must resolve minor disturbances and short-term ramping effects. Hourly balancing uses forecasts to meet loads at peak and off-peak demand which creates the daily shapes of energy consumption.

In the long-term, seasonal patterns affect renewable energy generation, the consumption of natural gas, and end-user power consumption. Naturally, these many structural and behavioral changes require technical solutions that are responsive at multiple timescales and can be applied to the grid periphery. Furthermore, these technical solutions will need to be supported by appropriately designed technology, policies, and regulations.

1.3 eIoT as an Energy-Management Solution

This work advocates the "energy internet of things " (eIoT) as a promising technical solution to the challenges presented above. The eIoT is one application of the internet of things (IoT). The IoT term was first used in 1999 by Kevin Ashton [115] and later became an integral part [116] of a global research consortium called the Auto ID Centre [116] that included the Massachusetts Institute of Technology (MIT), the University of Cambridge, ETH Zurich, Fudan University, Keio University, and Korea Advanced Institute of Science and Technology (KAIST). It is a technology that has expanded the use of communication technologies namely; over the internet, from user-to-user interaction to device-to-device interaction [117]. The adoption of the IoT has been supported by business efforts, such as the establishment of the Internet Protocol for Smart Objects (IPSO) Alliance in 2008, and technological advancements, such as the launch of Internet Protocol version 6 (IPv6) in 2011 [117]. Internet technologies with IoT have enabled growth in industry, especially in home automation and supply chains [117]. As a way to connect humans, computers, and devices, IoT presents itself as a key enabling technology of new energy-management approaches.

From the beginning, decentralized supply-chain management was an integral part of the IoT vision [5–13, 118]. The idea of was that the IoT provided unprecedented visibility of shop floor and supply-chain operations. Each piece of raw material, work in progress, or final product could be on tracked in near real time through the control loop captured by Fig. 1.8. When this information is relayed to manufacturing

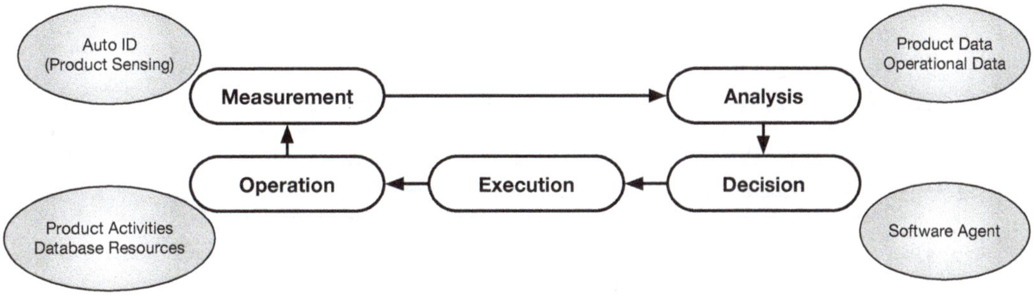

Fig. 1.8 A closed-loop control framework for production systems with intelligent products [5–13]

execution systems and enterprise information systems, it could be used to support reactive and proactive decision-making on how to best manage production systems and their associated supply chains.

Next-generation production systems [119–121] such as Industrie 4.0 advocated for the concept of "intelligent products" [8, 122, 123] that used "product agents" [124–131] that negotiated in real time with supply-chain resources to make it to their final customer. The presence of an embedded product sensor (e.g., a radio-frequency identification (RFID) tag) enabled this new paradigm in industrial control systems.

eIoT emerges when the vision of IoT described above is applied to "energy things." In other words, it forms a "digital energy network" [132] where IoT technology is integrated into the smart grid as a full supply chain that includes centralized generation, transmission, distribution, DERs, and customer premises. IoT enables opportunities for smart grid applications such as DG, DER, DR as well as TE. The distributed nature of these technologies makes them ill-suited for the hierarchical and centralized systems as is typically found in conventional bulk power systems.

The decentralization of the energy system requires device-to-device connectivity so as to achieve distributed energy management. Eventually, the number of devices (things) that connect to the periphery of the power system is expected to grow significantly. In the consumer market, the number of things that use electricity is far greater than the number of things connected to the internet. However, the number of internet-connected devices is rapidly increasing [133]. As electric loads become dynamic and responsive, it is imperative that the increasing number of "things" that connect to the grid are managed through faster, real-time communications and control.

When the concept of decentralized IoT-based supply-chain management is applied to "energy things," it has the potential to become a powerful energy-management solution that not only reaches the grid periphery but also addresses dynamics at multiple timescales. IoT can manage end-point devices with real-time communications and control, and achieves monitoring, tracking, management, and location identification through protocol-based communications and data exchanges [133]. Smart devices (RFID tags, sensors, actuators, etc.) connect via communication networks (cellular networks, ZigBee, WiFi, etc.) to decision-making entities and actuators [133]. The process forms an IoT-enabled control loop that can be used to monitor the equipment state of devices, collect information for analysis, and

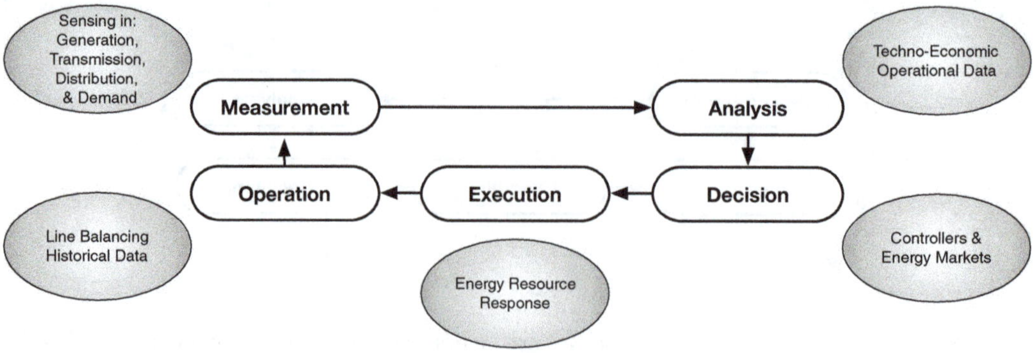

Fig. 1.9 A closed-loop framework for electrical power system management

control the smart grid for a variety of applications [133] (Fig. 1.9). For example, TE is the realization of a control loop interacting with market information, two-way communication networks, and real-time pricing mechanisms that incentivize the generation and consumption of electricity.

With the emergence of IoT, the technical development of the grid's infrastructure, the changing role of the grid's stakeholders, and the energy market development can all be advanced with real-time data. The ability to connect devices, create market signals, and influence generation and consumer behavior within an overarching energy-management framework is known as the *energy internet of things* (eIoT).

1.4 Scope and Perspective

The goal of this work is to provide a broad perspective of the implications of eIoT on the management and control of the electricity grid. This book offers a formal definition of the IoT within the context of the electricity supply and distribution control loop. It presents the growing demand for advanced and internet-enabled sensing and actuation devices for the generation and transmission system layers as well the distribution system layer. More importantly, it presents the changing roles of existing grid stakeholders as well as the gap in energy-management solutions that could potentially be filled by new stakeholders. Specifically, it recognizes a closer working relationship that may emerge through collaborations with telecommunication companies as new communication networks are adopted. Additionally, the book shows a convergence of cyber, physical, and economic frameworks as more eIoT devices seek to function and collaborate effectively. Finally, this work presents the role of TE as a core application of the eIoT control loop. Two TE use cases are presented to illustrate the changing nature of consumer interactions with utilities. This brings up the issue of how utilities are going to address the growing penetration of eIoT and DERs. Overall, the book presents the challenges, opportunities, and the transformative implications of eIoT on all the layers of the electricity supply and demand value chain.

1.5 Book Outline

To that end, the rest of this document is structured as follows:

Chapter 2 address the activation of the grid periphery.

- Section 2.1 recognizes that DERs will transform the nature of energy management at the grid periphery.
- Section 2.2 discusses some of the challenges presented by this transformation.
- Section 2.3 finally presents eIoT as a scalable energy-management solution for the activation of the grid periphery.

Chapter 3 focuses broadly on the development of eIoT within the energy infrastructure. This development is discussed in the context of a control loop.

- Section 3.1 presents the sensing and actuation in the transmission and distribution levels of the power grid. This section is discussed in four main categories:
  - Section 3.1.2 discusses sensing and actuation of primary variables in the transmission layer.
  - Section 3.1.3 addresses the sensing and actuation of secondary variables required for the reliable supply of solar, wind, and natural gas resources.
  - Section 3.1.4 introduces the sensing and actuation of primary variables in the distribution system focusing on key devices such as the smart meter.
  - Section 3.1.5 discusses sensing and actuation of secondary variables within the demand side, recognizing the role of automation, smart home devices, real-time demand-side data, and the challenge of integrating plug-in-electric vehicles.
- Section 3.2 presents the communication layer of the control loop recognizing that the current communication structure must evolve to deal with the heterogeneity of sensing and actuation devices. This evolution will occur within all layers of the energy system's jurisdictions.
  - Section 3.2 addresses the communication network for grid operators and utilities.
  - The shift from current grid communication networks to telecommunication networks is discussed in Sect. 3.2.3.
  - Section 3.2.4 addresses the growing demand for local area networks on the consumer side.
- Section 3.3 presents the need for distributed control algorithms to deal with the growing heterogeneity and number of control points in the electricity grid. This section examines the evolution of control algorithms and applications within multi-agent systems studies, game-theory approaches, and microgrid control.
- Section 3.4 discusses the changing architectural needs for the electricity grid and the need for standardization of cyber-physical/economic frameworks to enable interoperability of technologies.

- Section 3.5 examines the social implications of eIoT deployment both from the perspective of privacy concerns and eIoT cyber-security.

Chapter 4 presents TE as an overarching application of the eIoT control loop.

- Section 4.1 presents a broad definition of TE and offers a review of some of the current applications of the TE framework.
- Section 4.2 explores potential transformative impacts of TE in the energy system management. These impacts are summarized in two plausible eIoT use cases as potential transactive energy applications.
- Section 4.3 discusses the implications of eIoT for the future of electric utilities especially in North America, and finally,
- Section 4.4 considers the implications of eIoT for industrial, commercial, and residential consumers.

The book is concluded in Chap. 5 with a high-level discussion of the three main eIoT transformations in Sect. 5.1 and two major challenges and opportunities in Sect. 5.2. This chapter broadly reflects on the implications of eIoT advancement on the future of the electricity grid.

Chapter 2
eIoT Activates the Grid Periphery

Perhaps nowhere will the impact of the energy-management change drivers identified in Chap. 1 be felt more than at the grid's periphery. DG in the form of solar PV and small-scale wind will be joined by a plethora of internet-enabled appliances and devices to transform the grid's periphery to one with two-way flows of power and information [45, 46]. This transformation presents a daunting technical challenge. Not only are there tens of millions of devices at the leaves of the grid's radial structure, these devices are relatively small and require new innovations in sensing, communication, control, and actuation.

This chapter first describes this transformation in Sect. 2.1. Section 2.2 describes the challenge of activating the grid's periphery. Finally, Sect. 2.3 describes how eIoT can potentially be deployed as a scalable energy-management solution.

2.1 Change Drivers Will Transform Energy Management at the Grid Periphery

The installation of DG in the form of solar PV and small-scale wind causes two-way flows of power and information at the grid periphery. The change drivers discussed in Sect. 1 directly and indirectly incentivize growth in renewable energy generation. Renewable energy is, by nature, decentralized, and the deployment of small-scale power generation is increasing in industrial, commercial, and residential applications [134]. For example, the installations of solar PV systems in the USA nearly doubled from 2014 to 2016 [134]. Generation at the grid periphery introduces a power flow inward, or upward, towards the transmission system in addition to the normal outward power flow to consumers.

As the generation at the periphery of the grid continues to grow, energy-management systems must adjust from a "top-down" hierarchical structure of communication and control to one that is more dynamic and distributed [135]. The variable nature of renewable energy resources (for example, solar PV and wind)

S. O. Muhanji et al., *eIoT*, https://doi.org/10.1007/978-3-030-10427-6_2

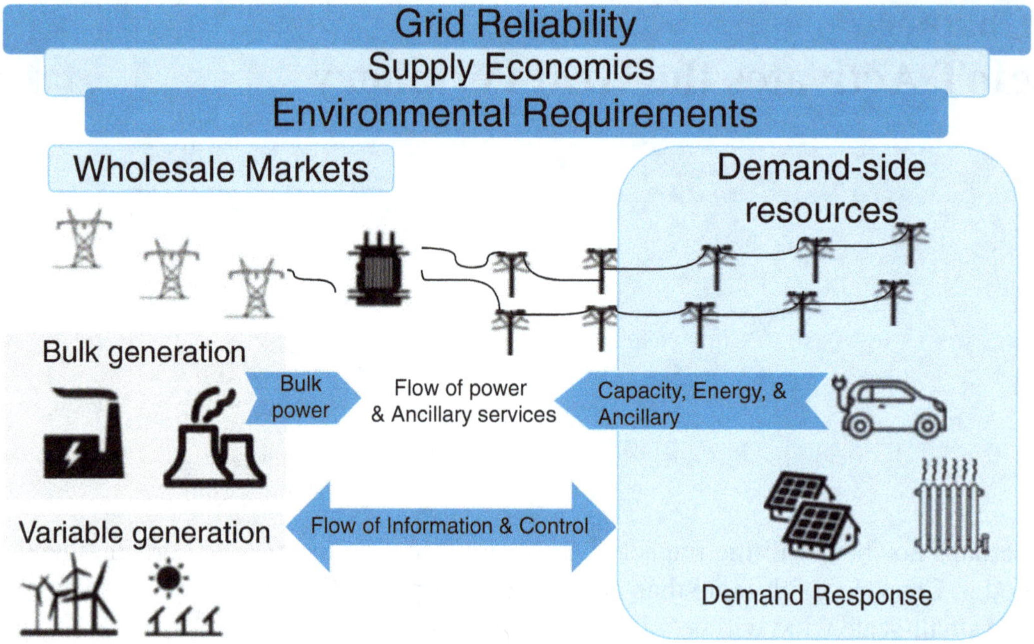

Fig. 2.1 A grid periphery activated by variable generation and demand response (adapted from [14])

means that in order to achieve sustainability, data acquisition, and new networks to monitor real-time power flows are imperative [136]. This is best illustrated in Fig. 2.1 which shows the need for two-way flow of information and control between the grid generation and transmission system and the grid periphery with a large penetration of distributed generation and demand response.

In addition to DG, a plethora of internet-enabled appliances and devices further reinforce the presence of two-way flows of power and information at the grid periphery. The demand side provides devices for controlling the balance of power consumption and generation through real-time demand response. High penetration rates of renewable energy motivate the need for real-time demand response; furthermore, deregulation and increased consumer participation is achieved with active economic real-time demand response. IoT devices, at the periphery, such as electrified vehicles (EVs); electricity storage in industry, commercial buildings, and residences; and smart devices in the home have created a new demand-side network of devices that requires the grid to become more dynamic as device interactions increase [97].

With drivers to incorporate DER and DR programs, smart grid technologies will enable end users to actively manage their electric loads according to price incentives. This active balancing of power at the grid periphery can shift in real time from positive (due to excess DG) to negative (due to modulated/controlled/incentivized) demand response. Internet-enabled appliances and devices in the grid periphery must be monitored and controlled in order to take advantage of real-time shifts in economic demand response. Bidirectional information flow sends pricing signals to the devices, while device information is sent to the controller. Where feedback loops are physical rather than economic, these devices can also potentially provide ancillary services in response to operational signals, for example, grid frequency, voltage, and line congestion.

The need to monitor and control two-way flows of power with two-way flows of information emphasizes the role of data gathering in the power grid. Data are needed to make accurate control decisions in the grid's increasingly flexible and fast-paced environment. Utilities are deploying more devices to collect more data of increasing diversity. The global number of devices being managed by utility companies is projected to grow from 485 million in 2013 to approximately 1.53 billion in 2020 [97]. Improved grid monitoring and control involves increasing the quantity of field distribution automation devices, field monitoring devices, substation monitoring and control, and interconnections and monitoring of independent power producers (IPPs) [137].

Also, future utility investments are expected to develop smart metering infrastructure across industrial, commercial, residential, transformer, and field meters [137]. Each application should accommodate a utility's business model and the network's specifications. For example, field distribution automation devices include remote monitoring and control of distribution reclosers, switches, voltage regulators, and capacitor banks that must be united under a common communication network [137]. All of these devices produce data at regular intervals, although there is a shift towards real-time data streaming. For instance, some smart sensor systems produce large streams of data from thousands of sensors, which—without appropriate planning and design—have the potential to overload system operators [138]. Due to the growing magnitude of deployed devices, and the use of proprietary and non-proprietary solutions, the monitoring devices on the grid produce increasingly heterogeneous data [139]. More devices, recording ever-more diverse measurements, create a thorough monitoring environment that has the potential to improve power system operations with new self-healing and reconfiguration capabilities. Granular data will also shift the grid from load-following to load-shaping energy management [3].

In order to support the two-way flows of information in the power grid, new networks are necessary. Many smart devices use applications that depend on data sets distributed across many devices. Furthermore, this information is often relayed to centralized centers for further storage, processing, and decision-making [140]. Multiple types of networks are required to co-exist. Although the supervisory control and data acquisition (SCADA) system gives utilities limited control of their upstream functions, the distribution network is insufficiently monitored and controlled [141].

As a solution, distribution-management solutions are expected to integrate with upstream SCADA as well as interoperate with the complex multitude of downstream network-enabled devices. In a survey sent to over 300 members of the Institute of Electrical and Electronics Engineers (IEEE) power energy system (PES) distribution-management system (DMS) task force, which comprises 76% utilities, about 72% of responders noted that SCADA facilities would be an integral part in distribution-management systems (DMS) [142]. Over 80% of survey participants also responded that more than one mechanism is necessary to handle DMS data acquisition and control requirements [142]. This is because SCADA's centralized and hierarchical structure is ill-suited for the developments in information and communication technology at the grid's periphery.

Because SCADA is a utility-purchased software that monitors hardware in the electricity infrastructure [143], consumer-owned smart devices are out of the realm of SCADA control. Therefore, consumer devices require either their own local area network (LAN) or access to a common network such as the internet. For example, a private solution-specified network may include machine-to-machine (M2M) systems that remotely read customer energy consumption and interface with power grid communications [97]. The IoT can further enhance the operational capabilities of M2M systems by connecting several such systems together [97]. Naturally, interoperability of the emerging networks is crucial. However, open network access raises privacy and security concerns. Cyber-security efforts must be directed towards individual devices as well as the communication channels between them. With many networks existing beyond the scope of the utility, these efforts are ever-more integral to the physical security of the grid.

As two-way flows of power and information become common place at the grid periphery, new energy market structures can evolve from their current hierarchy. The integration of renewable energy into market operations requires new measurements, measurement devices, and market information to ensure efficient and equitable operation [101]. As renewable energy and active demand-side resources become more prevalent, the grid's periphery will become not just a source of power, but also a place for diversified market activities [97]. As new market agents appear, they will require real-time measurements for market surveillance and contract compliance [101]. More specifically, DER incentives rely on bidirectional price and consumption data to be effective [144].

Grid and meter data can support the efficacy of these market mechanisms at both the wholesale and local levels. Furthermore, such data can help shape the development of monetized efficiency services based upon the real-time behaviors of residential, industrial, and commercial customers [97]. These trends, taken in the context of deregulation, encourage the participation of non-traditional parties [97]. DG, in particular, has the potential for large-scale market disruption. It is uncertain how the structure of energy markets will change as energy consumers evolve into prosumers [145].

2.2 The Challenge of Activating the Grid Periphery

The transformation of the grid periphery is a daunting technical challenge because it is characterized by millions of small devices; all of which need to be coordinated to achieve high-level technical and economic energy-management objectives. For example, actively shaping the load profile when it is composed of so many devices is a great challenge as it requires precision control, accurate forecasts, and flexible resources. Such a grid transformation poses integration challenges in operations as well as in the fiscal and strategic planning of distributed resources. Sensing equipment must improve to support demand-side management, and system planning requires cheaper devices that can be deployed at scale. In addition to extending sensing and control capabilities in the distribution system, other challenges in periphery management include inflexible loads.

The technical challenges of integrating the grid with peripheral devices in DR solutions, all through a consistent regulatory and economic framework, are staggering. The ongoing interconnection of the electric power system requires foresight and planning on the part of operators as well as regulators. All the while, the grid needs to be in full operation at its usual level of reliability and security.

Not only will the transformation of the grid periphery be complicated by their large number but also by their tremendous heterogeneity. This means that coordination and control algorithms must account for a wide variety of devices each with their own device-specific behaviors. "The future electric system will include a large network of devices that are not only passive loads, as most endpoints today are, but devices that can generate, sense, communicate, compute, and respond. In this context, intelligence will be embedded everywhere, from EVs and smart appliances to inverters and storage devices, from homes to microgrids to substations" [146].

Independent actors at the grid periphery are expected to add tens of millions of devices with different sizes, consumption patterns, time scales, and with different control and economic capabilities [146]. Such DERs (devices) include both generation and consumption. On the generation side, generation can be derived from wind energy systems, photovoltaic cells, microturbines, fuel cells, solar dishes, gas turbines, diesel engines, and gas-fired internal combustion engines [147]. Demand-side resources would include smart appliances, EVs, water heaters, air-conditioners, and energy storage in homes, buildings, and factories [148]. DERs also make use of power electronic interfaces so as to connect flexibly to the grid [147]. The centralized control of such devices is limited to hundreds or even a few thousand monitoring and control points.

As such, the distribution system is ill-equipped to control and coordinate the millions of homes, buildings, and factories with their associated energy devices [148]. Each customer and device has the potential to independently and dynamically interact with grid operations and markets. Such cases would require the implementation of complex algorithms for monitoring and control [146].

Due to the small size of devices and their increasingly complex interactions, the distribution system needs to be controlled with even more precision. Power system performance, control and daily operation use various mathematical models that need accurate generation, transmission, and distribution parameters in order to run [149]. It is very difficult to control and coordinate a large number of devices so that they achieve positive global objectives, especially when distribution monitoring is inadequate.

Such a multi-objective system coordination problem, that is, factoring not only improved system quality, security, customer service, and economics, requires more effective and robust control strategies [149]. Evaluating these different control options opens the question of whether the control architecture should exhibit hierarchy, heterarchy, or aspects of both. In the hierarchical system, linked aggregation points feed to a centralized control station. Aggregation is expected to be used in short-ranged sensor networks and connecting M2M networks with other technologies [150]. However, a comprehensive aggregation strategy is not clear. In heterarchy, control is distributed among centers with separated functions.

Present-day control centers are progressively characterized by separated control systems, energy-management models, data models, and middleware-based distributed energy-management system (EMS) and distribution-management system (DMS) applications [101]. Distribution control algorithms allow for scalability at pace with the growth of consumer nodes, but many suitable algorithms have yet to be developed. Most likely, the grid requires a mixture of aggregation and distribution philosophies to meet its diverse objectives.

To further complicate matters, the distribution system and grid periphery, unlike the transmission system, have not been traditionally monitored or controlled. Traditional, centralized control depends on independent system operator (ISO) supervision with the participation of large generators and load-serving entities. ISOs, however, cannot view the system past substations [151]. Essentially blind, operators are concerned about renewable generation at the periphery [148, 151]. ISOs currently aggregate variable net load at the transmission substation, which results in uncertainty that must be counterbalanced by expensive and inefficient operations, such as larger transmission and reserve capacity acquisition by the ISO and power providers [151].

Consequently, the activation of the grid periphery to include full control loops of sensing, decision-making, and actuation requires significant technology development and implementation. DERs must be visible and controllable by grid operators and planners in order to secure reliability and enhance economic efficiency. Such integration needs a framework for transmission, distribution, and demand-side resources that includes new analysis tools, visualization capabilities, and communications, and control methods [144]. Naturally, any effective strategy has to assume that there will be a migration from traditional passive devices to an ever-increasing but gradual penetration of network-enabled devices.

As more DG and network-enabled devices are integrated into power grid operations, utilities and grid operators are less able to accurately predict the stochastic net load profile. Since the inception of the electric grid, consumers have dictated the quantity of power that has been sourced by controllable generation. The design of the electric system was built on this paradigm; it was not intended for substantial amounts of uncontrollable generation, such as variable renewable energy [152]. In today's grid, operators turn on generators to meet a prediction of aggregated consumer demand. However, renewable energy's dispatchability (ability to dispatch to accurately meet demand) remains largely uncontrollable, and its predictability can change due to weather conditions and site-specific conditions [152, 153].

Past:

Generation/Supply	Load/Demand
Thermal Units: (Few, Well-Controlled, Dispatchable Resources)	Conventional Loads: (Fairly Slow Moving, Highly Predictable, Always Served)

Future:

	Generation/Supply	Load/Demand
Well-Controlled & Dispatchable	Thermal Units: (Potential Erosion of Capacity Factor) ↓	Demand Side Management: (Requires new control & market design)
Stochastic/ Forecasted	Solar & Wind Generation: (Variability can cause unmanaged grid imbalances) ↑	Conventional Loads: (Continuing source of variability & uncertainty)

Fig. 2.2 A future smart grid with stochastic and controllable supply- and demand-side resources [15–17]

As a result, forecast errors are expected to increase. Prediction models need to be individually developed per site, since local characteristics influence renewable power generation [152]. Utilities may develop such prediction models for large-scale renewable generation, but it is impractical to invent a separate model for each residential and small-scale distributed generator [152]. Referencing Fig. 2.2, the increasing penetration of variable energy is analogous to shifting from controllable loads to stochastic loads, but operator management of the system at large does not change as quickly. Forecast error is a long-standing operational challenge that will continue to grow as the penetration of renewable energy generation increases. In the immediate future, operation and control of demand-side resources must be precise in controlling set points of frequency, voltage, and line flows. Furthermore, these set points must be responsive to the errors propagated by inaccurate forecasting.

While the need for accurate forecasting in grid operations is ever-increasing, cost barriers remain to the implementation of advanced monitoring. Equipment expenses and other implementation objectives combat pressures for heavy monitoring in the grid. Conventional monitoring and diagnostic systems require expensive wiring and regular maintenance [154]. In contrast, wireless sensor networks (WSNs) have been pursued for their low cost, rapid deployment, and flexibility [154]. To deploy at scale, utilities maximize the per unit investment cost of sensing. For example, a fifty-dollar sensor on a 50-mW unit is far more valuable than the same cost sensor on a 50-kW unit. Such costs act as barrier to entry despite market deregulation [155].

Centralized generators often do not support investment in distribution monitoring systems, not just because of their costs, but also because they shift market power to DERs [155]. However, sensor technology developers are actively driving down the price of sensors for their widespread adoption. For example, the Auto-ID Center—the organization accredited with the term "Internet of Things"—set a goal of decreasing the cost of RFID tags from upwards of $0.50 to as low as $0.05 per tag [156]. Lower costs must come from new technologies and methods and cannot depend on simple economies of scale [156].

Finally, it is important to recognize that the control and coordination of demand-side resources is fundamentally more complex than supply-side resources. Besides operational challenges, short-term and long-term consumer behaviors will need to be altered through DER management and incentivized DR programs [46]. The ultimate objective of DR is to alter demand so as to enhance grid reliability and economic efficiency [46]. Nevertheless, it is complicated by the inflexibility and time-varying economic utility of loads. While supply-side management exists solely to serve demand, demand-side management (DSM) primarily supports a non-electrical activity, such as driving a motor or heating a building. Any behavioral shift (by DR programs) to support the reliable operation of the power grid is often at odds with the original intention of electricity consumption. Furthermore, it is important to recognize that a consumer's preference for electric consumption is time varying and "meddling" with service may lead to discomfort [46].

Fundamentally speaking, economic utility depends on the application of electric consumption. The value delivered by 1 kW of electricity for one purpose is not the same as the value delivered by another kW for another purpose, even if the kilowatt is consumed by the same customer! For instance, a manufacturing plant using 10 kW gets much more value when the electricity is consumed by a machine on the shop floor than by the back office. Uncertain economic utility and imperfect behavioral response make the control and coordination of demand-side resources particularly difficult.

2.3 Deploying eIoT as a Scalable Energy Management Solution

This work argues that the challenges of activating the grid periphery, described in the Sect. 2.2, may be addressed by deploying eIoT as a scalable energy-management solution. In essence, the energy-management challenges described in the previous section may be viewed as a control loop where dispatchable devices, whether they are traditional large-scale centralized generators or millions of small-scale internet-enabled devices, must meet the three power system control objectives of balanced operation, line congestion management, and voltage control. These objectives can be achieved despite the presence of disturbances such as customer load or variable energy generation from solar PV and wind resources.

Fortunately, eIoT is fundamentally a control loop consisting of small-scale sensing technologies, wireless and wired communication technologies, distributed control algorithms, and remotely controlled actuators. And yet, despite eIoT having all of the components of a scalable energy-management control loop, the challenge is to continue to integrate more of these technologies in such a fashion that the control objectives are achieved well into the future. Chapter 3 details the development of eIoT technologies in terms of their role in a control loop.

Chapter 3
The Development of IoT Within Energy Infrastructure

The development of IoT within the energy infrastructure is best seen as a control loop. The control loop is composed of four functions: a physical process (such as the generation, transmission, or consumption of electricity), its measurement, decision making, and actuation. This control structure is shown in Fig. 3.1 where a sensor takes measurements of the states and outputs of a physical system. Wireless and wired communications are used to pass this information between the physical layer and other informatic components. This information is used to make decisions either independently in a decentralized fashion or in coordination with the informatic components of other devices. Decisions are sent back down to network-enabled actuators for implementation.

In some cases, this control loop acts in near real-time; in other cases, some of the information is used as part of predictive applications that facilitate decisions at a longer timescale. Control algorithms implemented at different layers of this control loop enable the control of individual devices as well as the coordination of smart grid devices that make up other parts of eIoT. Given the connectivity between the functions of this control loop, its successful implementation requires architectures and standards that ensure interoperability between eIoT technologies.

This chapter serves to summarize the most recent developments of IoT within the energy infrastructure. The discussion proceeds bottom-up by classifying these developments according to the generic control structure shown in Fig. 3.1.

- Section 3.1 discusses some of the state of the art in network-enabled physical devices, whether they are network-enabled sensors or actuators in the control loop.
- Section 3.2 focuses on the communication networks that send and receive data to and from these devices.
- Section 3.3 discusses advancements in distributed control algorithms to coordinate the techno-economic performance.

S. O. Muhanji et al., *eIoT*, https://doi.org/10.1007/978-3-030-10427-6_3

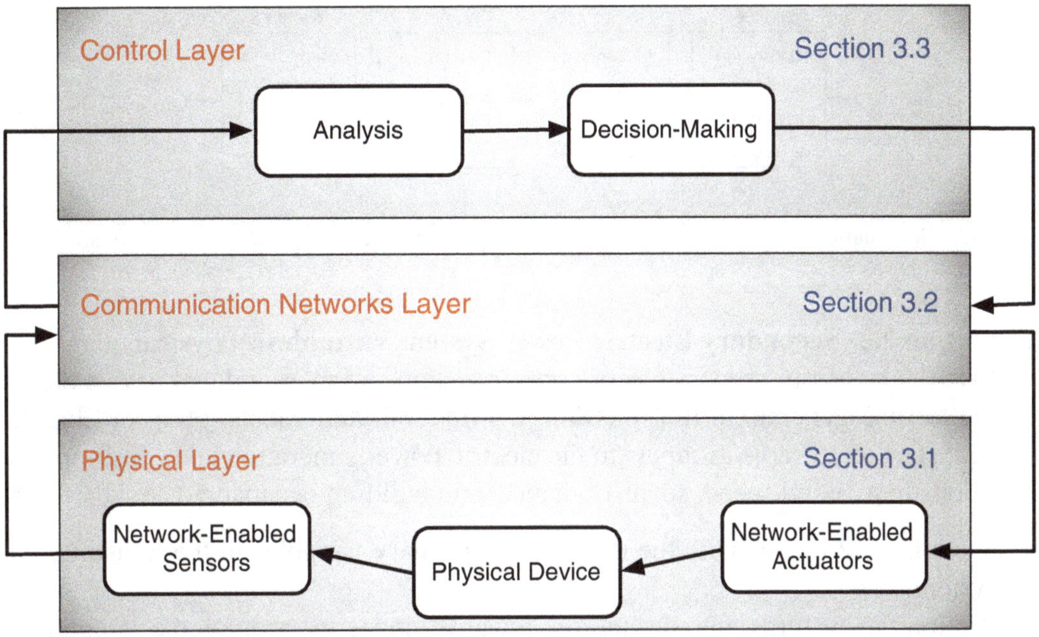

Fig. 3.1 The development of IoT within energy infrastructure as networked control loop

The chapter concludes with two discussions of a cross-cutting nature:

- Section 3.4 addresses the importance of control architectures and standards in the development of eIoT technologies.
- Section 3.5 addresses the security and privacy concerns that emerge from the development of eIoT technologies.

3.1 Network-Enabled Physical Devices: Sensors and Actuators

3.1.1 Network-Enabled Physical Devices: Overview

In many ways, the development of network-enabled physical devices forms the heart of eIoT implementation. As such, this section provides a broad review of these technical developments taking into consideration their tremendous heterogeneity and relative placement within the electric power system. Figure 3.2 provides a schematic overview of the section making sure to distinguish between the measurement and actuation of primary and secondary electric power system variables.

Definition 3.1 (Primary Electric Power System Variables) Physical quantities that describe the physical behavior of electric systems. They are voltage and current magnitudes and phase angles, active power, reactive power, magnetic flux, and electrical charge. ■

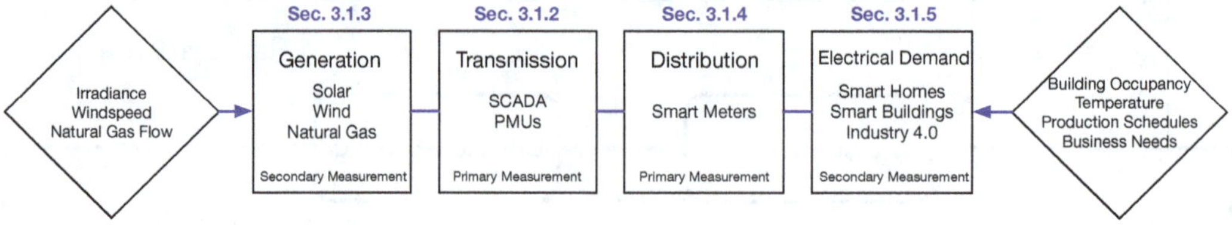

Fig. 3.2 Schematic overview of Sect. 3.1 on network-enabled physical devices: sensors and actuators

Definition 3.2 (Secondary Electric Power System Variables) Physical quantities that are distinct from primary electric power system variables and that have a direct impact on the generation, transmission, distribution, and consumption of electric power. They often serve as inputs to the electric power generation and consumption functions (e.g., wind speed, solar irradiance, and building occupancy). ■

- Section 3.1.2 begins with the (traditional) primary variables in the transmission system.
- Section 3.1.3 turns the discussion towards concerns around the secondary variables associated with wind, solar, and natural gas generation.
- Section 3.1.4 returns to the primary variables in the distribution system so as to address smart meters and other "grid modernization" technologies.
- Section 3.1.5 discusses smart homes, industry, and transportation in the context of demand-side secondary variables. Each of these sections addresses network-enabled sensors and actuators.

Sensing technology plays an indispensable role in providing *situational awareness* within an eIoT control loop that activates the grid periphery. As such, sensors exist at the periphery of a communication network to relay data and information from the physical grid to a control or decision-making center. Given the tremendous heterogeneity in the number, type, and input of physical eIoT devices, the measurement role of network-enabled sensing technologies increases immensely. Fortunately, there has been significant innovation in sensing technologies to accommodate these needs. Such advancements include miniaturization, wireless data transfer, and decreasing implementation costs. Miniaturization technologies have enabled monitoring of household devices where it was previously infeasible to collect data. Noninvasive wireless technologies have reduced implementation costs by forgoing wired installation. These two factors have made sensors increasingly ubiquitous in electric grid applications.

Although network-enabled sensors vary in design and location within the power system, they have a commonality of function that is fundamental to measurement within the control loop. At a basic level, a sensor is composed of a sensing unit, a processing unit, a transceiver unit, and a power unit [138]. Depending on its function, a sensor component must balance various design aspects such as power consumption, memory allocation, lifespan, and cost [138]. These trade-offs lead to a heterogeneity in sensor operations such as data collection intervals, wired or wireless communication, type of power source, and their connection to other devices. Furthermore, and as mentioned in Sect. 2.2, the need for precise control and accurate net load forecast also drives the deployment of a greater heterogeneity of sensors [138]. Here, the distinction between primary and secondary variables becomes important. Traditional primary variables have often been measured first due to physical and monetary constraints [157]. However, the need to better characterize variable energy, energy storage, and demand-side resources has led to the development of secondary measurement applications as well. These additional measurements improve situational awareness because they show the underlying causes for the supply and demand of electricity.

3.1.2 *Sensing and Actuation of Primary Variables in the Transmission System*

3.1.2.1 Network-Enabled Sensors: SCADA and PMUs

The development of monitoring and sensing technologies began in the transmission system in response to the Northeast Blackout of 1965 [158, 159]. It was found that as the North American power system became ever-more connected it was necessary to deploy new sensing technology so as to gain greater *situational awareness* of the transmission system as a whole. As shown in Fig. 3.3, a tremendous heterogeneity of sensors is deployed in the transmission system where they are used in transmission lines and substations to monitor "traditional" variables directly

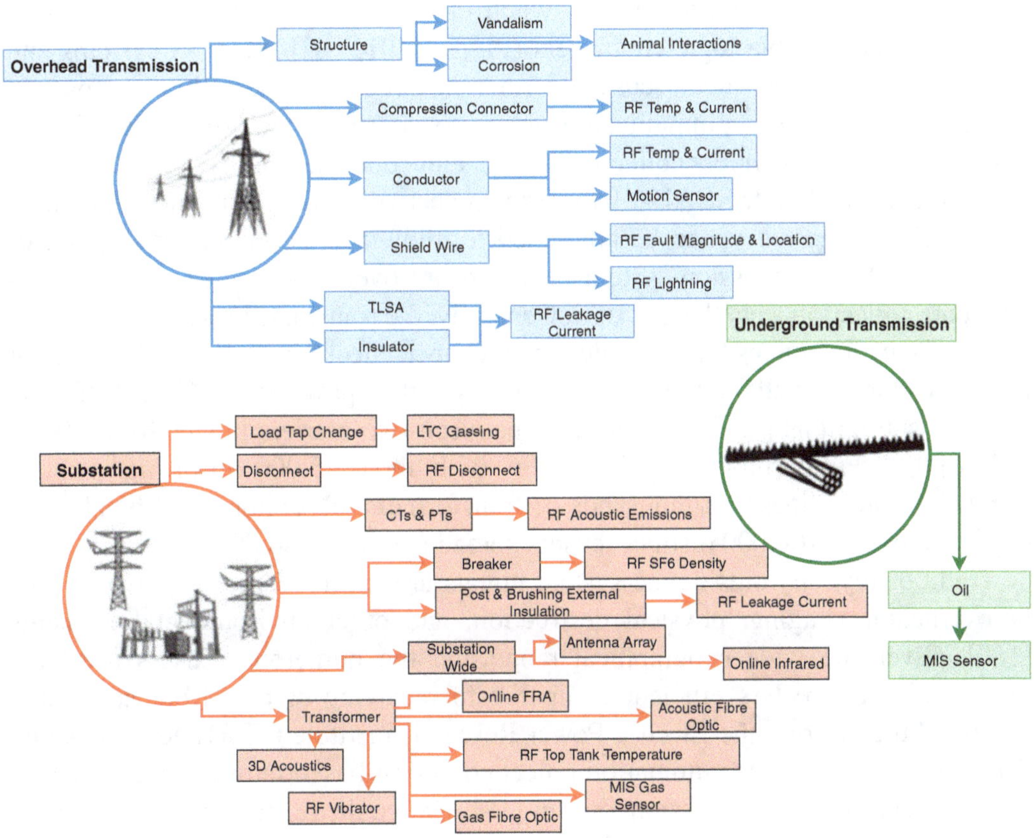

Fig. 3.3 Sensor technologies in transmission lines and substations (adapted from [18])

related to power quality, operations, and system limits. These variables are key to ensuring system stability and reliability and include voltage, current, their phase angles, active power, and reactive power. In the transmission system, line monitoring is achieved through sensors that measure voltage, detect faults, and conduct predictive maintenance [160].

Transmission sensors also help to monitor the physical condition of power supply equipment to improve safety, and determine when to deploy a workforce for repairs or outage prevention [18]. These sensors can be deployed in substations, in overhead lines, or in buried lines used for underground cable systems [18]. Sensors in the transmission system can also inform operational databases [18] to guide decision making that ensures system reliability. The reader is referred [18] for a deeper review of existing technologies.

The need for situational awareness also motivated the development of *sensor networks*. As is discussed in greater depth in Sect. 3.2, sensor networks are a collection of sensors tied to a modular communication network that bridge the gap between physical devices and decision-making points elsewhere [161]. These sensing networks are spatially distributed across the electric grid to form an interconnected monitoring and perception layer. The first and most prominent of such sensor networks is the SCADA system [19, 101, 162] shown in Fig. 3.4. SCADA is deployed in substations and distribution feeders where it is able to sense voltage, frequency, and power flows, and then send these measurements to centralized operations control centers. SCADA systems are also able to send remote signals to change generation levels, switch circuit breakers, and control devices through programmable logic controllers (PLCs) [101, 162]. SCADA systems and other sensor networks are discussed further in Sect. 3.2 where they are part of a larger discussion on communication networks. Further mention of the SCADA system in this section refers collectively to its embedded sensors.

Despite the elaborate SCADA-based sensing network in the transmission system, several challenges are yet to be addressed to allow for the effective adoption of eIoT. First, the transmission system is spread out over a wide area, making real-time data collection a challenge [163]. Generally, the transmission system is remote and deploying resources for scheduled maintenance checks is costly [164]. Many of the sensors are located on transmission carriers with approximately 60–125 carriers between substations [160]. The distance between two carriers ranges from 400 to 800 m [160]. Furthermore, a typical utility with about 25,000 km of high-voltage ($\geq$69 kV) power lines and thousands of transformers, capacitors, and breakers is expected to have 100,000 distinct sensors spread over a 20–80,000 km^2 area [138].

Traditionally, any outside-the-system threats are from weather (such as storms or overheating), aging, physical destruction, and other environmental elements [160]. Given the wide geographical range and the numerous sensors involved, manual checks are less efficient compared to receiving signals from automated sensors. Furthermore, the Electric Power Research Institute (EPRI) advocates that data communication and automation reflect condition-based rather than time-based management of the transmission system [18]. Probabilistic (rather than deterministic) methods for assessing risk in the transmission system can also be used to

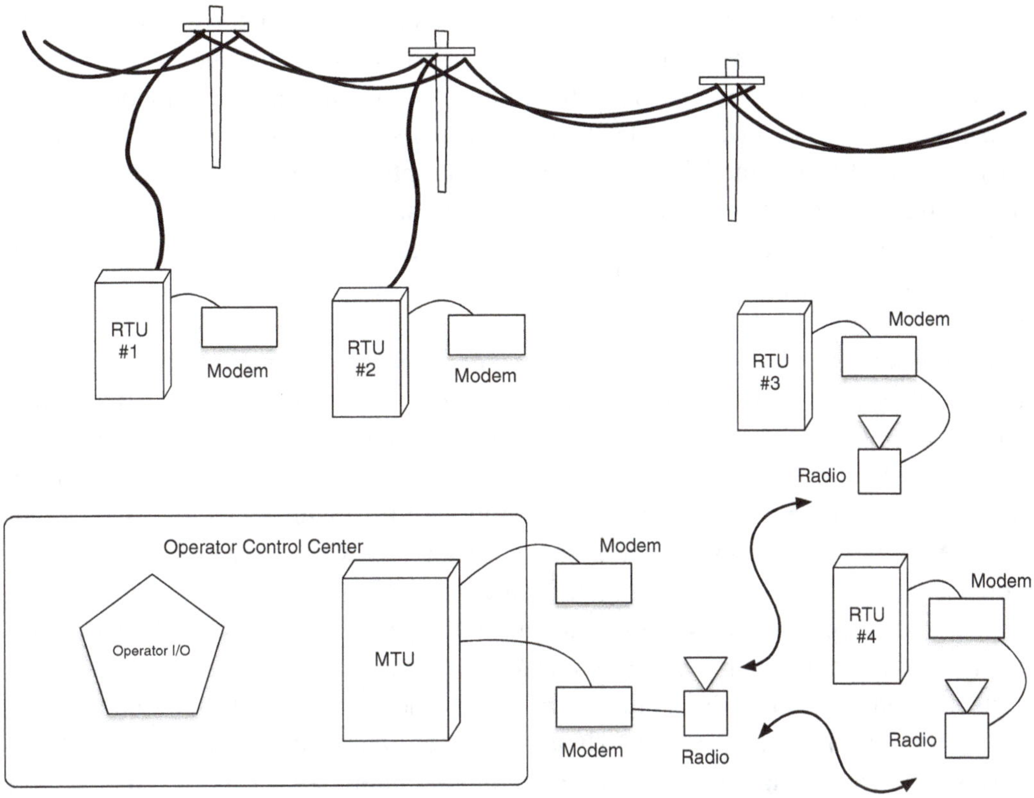

Fig. 3.4 SCADA as a network of remote terminal units (RTUs) connected to a master terminal unit (MTU) via modems and radios [19]

preemptively solve faults and address sub-optimal conditions [18]. In all cases, real-time data is needed to better monitor the conditions of the transmission system to ensure safety and reliability [138].

Second, the SCADA system, currently in place, cannot observe the dynamic phenomena in transient and small signal stability models [163]. SCADA has a relatively low sampling rate of 2–4 s, making dynamic state estimation over a wide area difficult [163]. Instead, SCADA data are often used in static state estimation algorithms [165–168] for manual decision making [169, 170]. Dynamic state estimation is further complicated by SCADA's lack of measurements with synchronized time stamps [163].

To address these issues, SCADA systems must be equipped with the ability to study temporal trends with finer resolution and synchronization [169]. These requirements imply better coordination and compatibility between SCADA terminals [163]. Such developments in wide-area measurements are set to enhance corrective actions against system-wide disturbances [171]. All in all, the electric grid must be updated with new sensors to enable the better gathering, transfer, and processing of measurement data [172].

Sourcing power for sensors can pose a major challenge to their deployment in sensor networks. The main energy intensive components in a typical sensor include microcontrollers, wireless interfaces, integrated circuits, voltage regulators, and memory storage devices. Nevertheless, this challenge can be overcome through the use of batteries or environmental power sourcing techniques [18]. A key factor in designing sensors for remote applications is ensuring sustainable energy consumption and supply. In order to minimize operation and maintenance costs, sensors must be designed in such a way that optimizes hardware and software energy use while taking advantage of energy harvesting opportunities from naturally occurring sources of energy such as thermal, solar, kinetic, and mechanical energy [138, 173]. Furthermore, some sensors can switch between a static "asleep" and a dynamic "awake" mode as needed.

In addition to such energy minimization techniques, designers must also optimize the use of passive components such as capacitors, resistors, and diodes to reduce leakage currents and switching frequencies [138]. Reducing the energy dependence of sensors on the electric power grid is of vital importance to prevent cascading failures between the physical electric grid and the informatic sensor network [174]. Such decoupling of the power grid's sensors from its physical power flows serves to increase the resilience of the two systems together [174].

These sensing challenges in the transmission system have motivated the deployment of phasor measurement units (PMUs) (that is, synchrophasors). Phasor measurements provide a dynamic perspective of the grid's operations because their faster sampling rates help capture dynamic system behavior [169, 170, 175–185]. PMUs measure voltage and current, and can calculate watts, vars, frequency, and phase angles 120 times per power-line cycle [163, 176]. Figure 3.5 shows the schematic of a PMU. This PMU data immediately enhances topology error correction, state estimation for robustness and accuracy [163], faster solution convergence, and enhanced observability [186]. Simulations and field experiences also suggest that PMUs can drastically improve the way the power system is monitored and controlled [186]. However, the installation of PMUs and their dependent solutions can be hindered by monetary constraints [186, 187]. A completely observable system requires a large number of PMUs which utilities usually install incrementally [187].

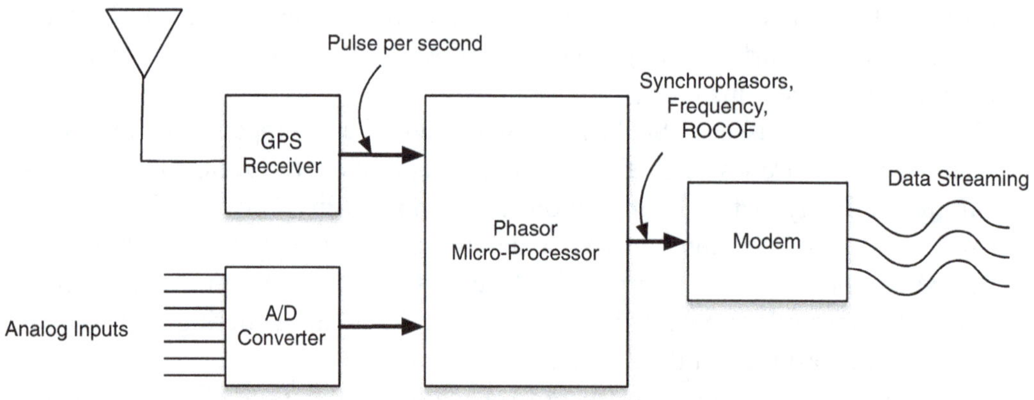

Fig. 3.5 Schematic of a Phasor measurement unit [20]

Recent studies have explored algorithms for optimal placement of PMUs to minimize the number of PMUs required to collect sufficient information [188–190]. PMU-based wide-area monitoring systems (WAMS) use the global position system (GPS) to synchronize PMU measurements [170]. Such synchronized measurements allow two quantities to be compared in the real-time analysis of grid conditions [186]. Through wide-area monitoring and synchronization, PMUs have made great strides in power system stability [170] which was often hindered by SCADA's slow state updates [191]. The implementation of synchrophasors has also allowed voltage and current data from diverse locations to be accurately time-stamped in order to assess system conditions in real-time [186]. Synchrophasors are also available in protection devices, but since requirements for protection devices are fairly restrictive, the full integration of synchrophasors into line protection is still debated [186]. The increasing application of synchrophasors in wide-area monitoring, protection and control systems, post-disturbance analyses, and system model validation has made these measurement tools invaluable [176, 187].

While the integration of PMUs into the transmission system will do much to enhance situational awareness in the transmission system, it is by no means sufficient for the grid as a whole. First, PMUs are primarily meant for applications in the transmission system and to a large extent are not feasible in the distribution system. They are even less appropriate for understanding customers' power consumption profiles. In that regard, the emergence of smart meters has fulfilled a much needed functionality. Second, PMUs only measure voltage and current phasors. As such, they are able to provide much needed insights into grid conditions but are not able to inform why these conditions exist. As the electric grid comes to depend more on interdependent infrastructure, weather conditions, and consumers' dynamic behavior, secondary measurements of these quantities become increasingly important. In that regard, sensors used in other sectors will have an indispensable role in taking secondary measurements.

3.1.2.2 Network-Enabled Actuators: AGC, AVR, and FACTS

In order to take full advantage of the heterogeneity of sensing and measurement technologies, a heterogeneity of actuation methods is also needed. Much like with sensing technologies, actuation technology has long been a part of power systems operations and control. Perhaps, the earliest remotely controlled actuator in the electric grid is automatic generation control (AGC) [192] which is used to maintain grid frequency in the face of fluctuating consumer load. In time, power system operations came to include automatic voltage regulation (AVR) [193, 194] to maintain voltage stability. Finally, a plethora of flexible alternating current transmission system (FACTS) [195] devices have been developed to address line congestion in addition to supporting AGC and AVR technologies.

AGC, formerly known as load-frequency control was established in the early 1950s [196] to adjust the power output of interconnected generators in order to meet variations in load (Fig. 3.6). Imbalances in real power generation and load

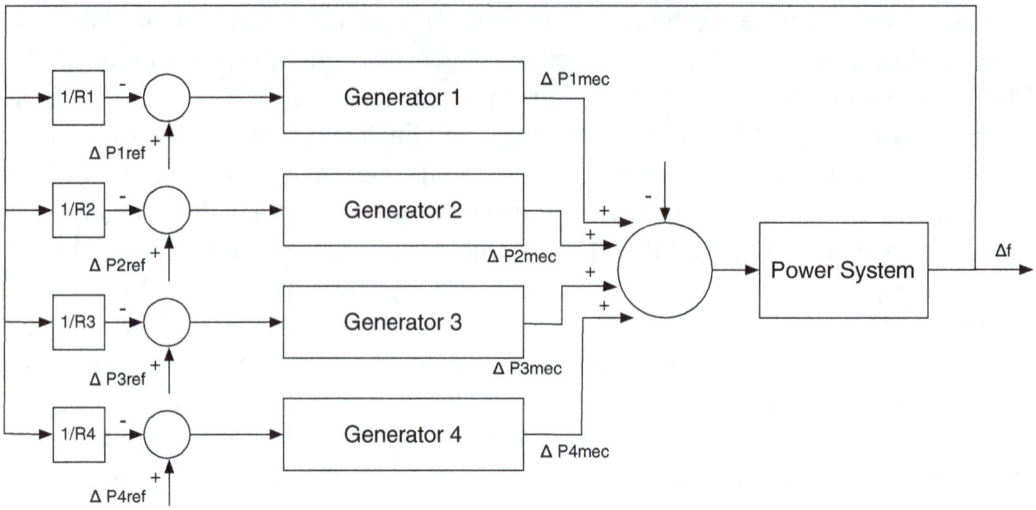

Fig. 3.6 Schematic of automatic generation control [20]

cause frequency fluctuations that could compromise the stability of the system. For a given control area, each energy control center aims to maintain zero area control error (ACE). ACE defines the difference between the net interchange power and the deviation in net frequency in megawatts (MWs) [196]. Controlling the ACE is the main role of AGC, and it is achieved through a mix of specialized control algorithms and automatic signals to generators. AGC achieves control of output generation by sending signals to generators every 4 s. The ability of generators to respond to these signals is governed by various characteristics of the generator, such as type of plant, fuel type, age of the unit, as well as operating point and operator actions [197]. In most cases, units under AGC tend to have faster ramping capabilities, such as fast start natural gas units.

As the electric grid becomes more and more interconnected, the AGC process has been complicated and research into distributed control algorithms for AGC is steadily underway [198]. (See Sect. 3.3 for further explanation.) AGC control has also become more decentralized with the Federal Energy Regulatory Commission (FERC) even allowing third-party AGC [199]. Such decentralized AGC is more likely to require advanced communication for any large-scale application to be considered feasible. Specifically, the current star-shaped communication architecture would need to change to a meshed one [172].

In addition to frequency regulation, voltage regulation is a key component in ensuring power stability. Voltage stability regulation has played a significant role in controlling the reactive power flow in the electric grid. The schematic of automatic voltage control is best captured by Fig. 3.7. In North America, voltage control is done at a local level although there is a possibility of expanding this to a regional level [172] where it has been successfully implemented in China and the UK. Voltage instability occurs when a condition in the system results in deficient reactive power. Currently, voltage instability analyses have relied heavily on contingency analysis to prevent conditions that could potentially result in deficient reactive power

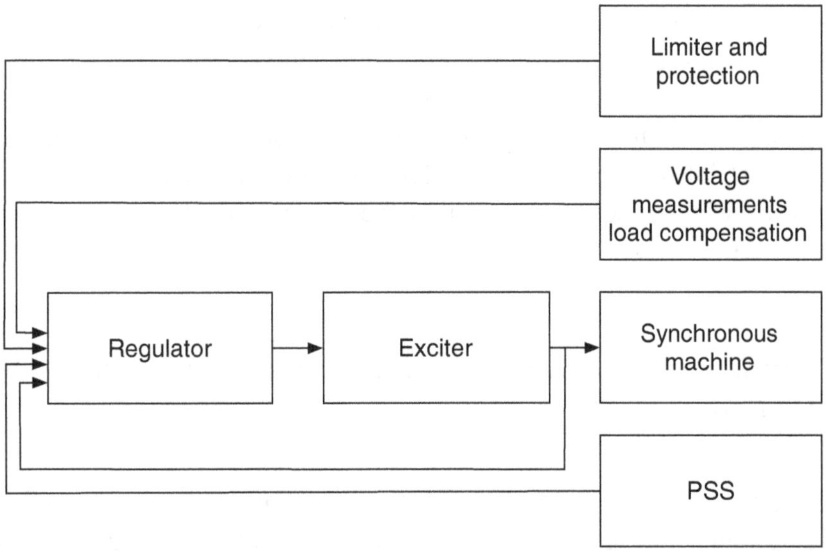

Fig. 3.7 Schematic of automatic voltage regulation [20]

[172]. This contingency analysis and prevention has been made possible by the use of automatic voltage regulators. With DERs, issues such as steady-state voltage spikes are likely to occur making the use of a single voltage regulator for multiple feeders infeasible [200]. Going forward, possible multi-agent approaches could be applied to provide more flexibility to the voltage regulation process [201].

The use of FACTS in power transmission has tremendously improved the amount of power that can be transported within the power grid. This has enhanced the stability of the grid in the face of increasing demand and variable generation capacity. FACTS devices can increase or decrease power flow in certain lines and respond to instability problems almost instantaneously. These devices have aided in power routing and have helped send power to areas that were previously insufficiently connected [202]. FACTS devices are a wide range of power electronic devices that are split into three categories depending on their switching technology: (1) mechanically switched, (2) thyristor switched, or (3) fast-switched [202]. They include but are not limited to: static synchronous compensator (STATCOM) and static VAR compensator (SVC) for voltage control, thyristor controlled phase shifting transformer (TCPST) for angle control, and thyristor controlled series compensator (TCSC) for impedance control [202]. SVC is an automated impedance matching device that switches in capacitor banks to bring up the voltage under lagging conditions and consumes VARs from the system under reactive conditions.

The SVC and TCSC represent what is commonly referred to as the first generation of FACTS devices [202]. A STATCOM is based on a power electronics voltage source converter and can act as a source or sink for reactive AC power as needed. This device is commonly used for voltage stability and belongs to the second generation of FACTS devices [202]. FACTS devices have played a key role in deregulated markets by helping to increase the load ability for power lines, reduce system losses, improve the stability of the system, reduce production costs, and control the flow of power in the network. These functions make FACTS devices indispensable as the electric grid becomes more interconnected and adopts eIoT. As eIoT develops even more, FACTS devices may need to become smarter so as to receive signals and regulate flow as necessary. Such facilities are particularly helpful in the control of DERs. The ability to connect to communication networks is also necessary for these devices to ensure that they communicate and work with other sensors and wireless devices.

3.1.3 Sensing and Actuation of Supply Side Secondary Variables

As mentioned earlier in the section, the deployment of variable energy, energy storage, and demand-side resources requires a greater understanding of their associated secondary variables. For example, the power injection and withdrawal of these resources depends on solar radiance, wind direction and speed, temperature, humidity, and rain [160]. Therefore, sensing and actuating these secondary variables enables the control of the supply and demand of electricity based on its root causes.

3.1.3.1 Networked-Enabled Sensors: Wind, Solar, and Natural Gas Resources

Perhaps the best way to appreciate the benefits of measuring secondary variables is by observing how IoT analogously enabled "smart manufacturing," which is defined as "the use of information and communications technology to integrate all aspects of manufacturing, from the device level to the supply chain level, for the purpose of achieving superior control and productivity [203]." Smart manufacturing implies the use of embedded sensors and devices that communicate with each other and other systems [203]. Through data gathering and sharing, these devices inform decision making and automation throughout the manufacturing network [203]. The system uses big data to improve, evaluate, and analyze operations, consumer interests, resource planning, and management systems via cloud-based tools [203].

Smart manufacturing involves a holistic approach where it tracks a product's life cycle from raw material, to factory, to end use [203]. Most important, smart manufacturing makes use of a distributed approach by ensuring that every entity in an organization has the necessary information, at the time it is needed, to make optimal contributions to the overall operation through informed, data-based decision making [203]. Systems such as *Industrie 4.0* advocated for the concept of "intelligent products," which used "product agents."

Furthermore, IoT has enabled greater supply chain integration both upstream and downstream of a given production system [119–121]. The information about incoming parts and services from upstream suppliers help streamline operations management decisions [8, 122, 123]. Similarly, the information about downstream demand allows production systems to manage when and where they need to deploy resources closer to real-time [124–131]. When the electric power system is viewed as a full supply chain, it can mirror smart manufacturing applications to extract the full value of eIoT.

In that regard, the reliable integration of solar and wind resources requires secondary measurement applications in the electric grid. Such measurements include *wind speed* and *solar irradiance*. This kind of secondary monitoring of weather-dependent variables is not entirely new to electric power systems. Hydrologists have been monitoring water flows and elevations to understand the potential for hydropower generation for decades [204]. Indeed, as concerns over global climate change and water availability rise, the *energy-water nexus* has received considerable attention [205–212, 212, 213, 213–225]. These works have investigated the availability of water for the energy infrastructure [217–225], the co-optimization of water and energy infrastructure [212, 213, 213–216], and the impacts of water consumption on the electric grid demand-side management [220, 226, 227].

However, solar and wind resources, unlike hydropower, are often called variable energy resources (VERs). They exhibit *intermittency* in that their power generation value is not entirely controllable. They also exhibit *uncertainty* in that their power generation value is not perfectly predictable [228–233]. In both cases, access to real-time secondary measurements of weather-based variables can greatly reduce the uncertainty they impose on electric power system operations [234, 235]. Furthermore, as solar and wind resources become more prevalent at the grid periphery as DG, concerns over voltage fluctuations, power quality, and system stability necessitate better forecasting [109].

Despite these similarities, solar and wind power generation requires distinct prediction and monitoring techniques. Solar PV monitoring is best served with effective short-term predictions of fluctuations in solar irradiance over short intra-day and intra-hourly timescales [109]. Such predictions when combined with the fixed parameters of the solar PV arrays (for example, size and efficiency), they can be used to calculate power generation values [109]. In most cases, forecasting techniques based purely on historical data are insufficient. Instead, many of the most promising approaches propose hybrid machine-learning techniques that combine historical data with real-time weather data [236].

Wind power generation also combines wind speed predictions with site-dependent variables such as surface landscape and weather conditions to accurately predict power output [236]. In both cases, solar and wind variability occurs on all timescales, from turbine control occurring from milliseconds to seconds to integrated wind-grid planning occurring from minutes to weeks [237–239]. Furthermore, wind and solar predictions quickly lose accuracy at longer timescales [232, 237, 240–244]. Consequently, a holistic approach to forecasting must address the many applications of power system operations and control [15]. These include reserves procurement and energy market optimizations such as unit commitment and economic dispatch [237, 245–250]. Advanced sensing technologies introduced through eIoT are expected to play a key role in obtaining and communicating raw data inputs to solar and wind prediction models.

Similar to VERs, even dispatchable resources such as natural gas can have variable supply chains that require secondary measurement to ensure reliable grid operation. The challenge of natural gas relative to other dispatchable power generation fuels is that its gaseous state requires purpose-built facilities for its storage. Coal and oil are often stockpiled at the input of power generation resources to ensure an effective ramping response to grid conditions. Natural gas, on the other hand, is fed by pipeline and has only limited storage capability in many geographical regions.

Therefore, the flow of natural gas is quite susceptible to pipeline capacity constraints. As the price of natural gas has fallen in recent years (in response to the expanded availability of shale gas), this susceptibility has only grown. Some ISOs now have over 50% of their power generation capacity come from natural gas units [251]. To ensure reliability, power grid operators must now coordinate their operations with natural gas operators to make certain that sufficient natural gas capacity is available for power generation [252].

And yet, coordinated operation of the natural gas and electric power systems requires a recognition of their inherent similarities and differences. The natural gas industry, like the electric industry, has undergone deregulation to encourage competitive markets [252–254]. The electric power system has wholesale energy markets that implement security-constrained unit commitment (SCUC) and security-constrained economic dispatch (SCED) decisions. They competitively clear 1 day ahead and every 5 min, respectively [253]. Meanwhile, natural gas supply contracts have durations from 1 day to 1 year [254]. This optimal supply mix of natural gas also compensates storage and not just supply and transmission (as is the case in electric power) [254]. Furthermore, natural gas is transported by shipment as liquefied natural gas or by pressure differences in a pipeline network as a gas [252]. In contrast, electricity has no such differentiation of material phase. Finally, the natural gas system has an entirely different set of organizations, regulations, and scopes of jurisdiction that further complicate coordination with the electric power system.

Nevertheless, the presence of deregulation and market forces now means that natural gas and electricity prices are often closely correlated [255]. This is especially true during particularly hot or cold days when both systems experience peak demand from heating, ventilation, and air conditioning (HVAC) units [253]. The challenge during these times is to design the control room operations and the markets for both commodities such that both infrastructures continue to operate reliably and cost-efficiently [252–263]. Naturally, these requirements further motivate the need for secondary measurement from eIoT.

3.1.3.2 Networked-Enabled Actuators: Wind and Solar Resources

The effect of VERs on power system stability and control is significant due to the intermittent nature of resources such as wind and solar. However, recent studies and applications are showing that these resources are not so variable after all. In fact, they can be used to provide ancillary services such as frequency and voltage regulation or "artificial inertia." Wind turbine generators have varying reactive power regulation capabilities, depending on the manufacturer. Types 1 and 2 wind turbines are based on induction generators and have no ability for voltage control. While types 3, 4, and 5 wind turbine generators have power electronic converters that allow them to control reactive power and regulate voltage [264].

Although Type 1 and 2 wind turbines cannot control voltage directly, they are usually fitted with power correction capacitors to maintain the reactive power output at a fixed set point [264, 265]. These voltage control capabilities can be used to regulate the voltage at the collector bus of the wind farm [264, 265]. A centralized controller would usually communicate with individual wind turbines directly to regulate their voltage. Presently, grid codes require wind power plants (WPPs) to have a specified reactive power capability (for example, 0.9 lagging to 0.9 leading), making reactive power capabilities fundamental to the design of WPPs [264, 265].

In recent years, the concept of "synthetic" or "artificial" inertia has been introduced as a potential application for frequency control. A study conducted on the New Zealand system explored a possible use of wind turbine generators for frequency regulation by providing a megawatt contribution within a small period of time [266]. The study also proposed the following activation mechanism to mimic the first frequency response produced by real inertia: (1) the activation must occur within 0.2 s after the frequency reaches 0.3 Hz lower than nominal, (2) the ramp rate of the output must be no less than 0.05 pu/s of the machine's total capacity in megawatts, (3) the output must be maintained for at least 6 s from activation, and (4) the machine must deactivate the artificial megawatt output once the frequency has returned to the nominal frequency [266]. With this activation technique, low inertia devices can contribute MWs towards a falling system frequency. Other studies have also proposed a mechanism of reprogramming power inverters connected to wind turbines to imitate "synchronized spinning masses" or synthetic inertia [267]. Hydro-Québec TransÉnergie was the first to adopt this application of synthetic inertia and the general response is good although not enough to sustain the growing penetration of wind [267]. As wind turbine designs advance to supply more inertia, they are increasingly viewed as contributors to system stability.

The nature of remotely controlled devices requires them to be self-sufficient and self-sustaining. Remote devices include power transmission line monitoring systems, sensors, backbone nodes, video cameras set up in the transmission lines and towers. Given their location, repair and maintenance of these devices is severely limited. As such, remote devices are constrained by battery capacity, processing ability, storage capacity, and bandwidth [161]. These devices are in need of remote sources of power although they can use power acquisition technology [161] to harvest their own power. In addition, these devices must be suited for varying environmental conditions and must be waterproof, dust-proof, anti-vibration, anti-electromagnetic, anti-high-temperature, and anti-low-temperature [161]. Data fusion technology has been suggested as an application that can be used to collect data more efficiently, and combine useful data for these remote devices [161].

As for solar PV actuation, smart inverters are seen as key components for the effective coordination of solar PV systems with other eIoT devices. Inverters play a key role in the intersection between the measurement and decision-making layer of the control loop. New developments in the field of power electronic devices and modern control strategies for inverters have provided numerous operation strategies for efficient management of the inverter-controlled systems. However, future inverter designs need to allow for modularity to ensure independent scalability of components especially when deploying them to distributed systems such as solar PV installations [268]. Modular inverter design is also key to fast and effective standardization [268].

With smart inverters, the integration of IoT devices with the direct current interfaces has become much easier [268]. For an inverter to be considered smart, it must have a digital architecture with the capability for two-way communication and a solid software infrastructure. The ability to send and receive messages quickly is imperative for effective eIoT deployment. Smart inverters must be capable of sending granular data to utilities, consumers, and other stakeholders quickly. This allows for faster and more efficient diagnosis of problems as well as maintenance [269]. For solar PV, smart inverters have a key role to play in improving system costs and performance as they provide high redundancy through distributed AC architecture [269]. Microinverters provide a PV system with the ability to provide ancillary services such as ramp rate control, power curtailment, fault ride-through, and voltage support through vars [269].

To fully develop and incorporate smart inverters to the grid, designers must work with utilities and regulators to meet the desired standards and regulatory requirements. The Underwriters Laboratory/American National Standards Institute (UL/ANSI) 1741 and IEEE 1547 standard groups together with the Smart Inverter Working Group (SIWG) are some of the groups that are working collaboratively towards advancing this technology [269].

3.1.4 *Sensing and Actuation of Primary Variables in the Distribution System*

As was discussed extensively in Chap. 2, the greatest transformation of the electric power grid will occur at the grid periphery. These include the integration of network-enabled sensors and actuators in *distributed* generation, distribution lines, and end-user power consumption. The discussion provided in Sect. 3.1.3, in many ways, already addressed the sensing and actuation of DG. Because solar PV and wind turbines are effectively scalable technologies, they may be integrated equally effectively in the transmission and distribution systems. Consequently, the conclusions of Sect. 3.1.3 are equally applicable here. This section now addresses the sensing and actuation of primary variables in the distribution system prior to addressing secondary variables in Sect. 3.1.5.

3.1.4.1 Network-Enabled Sensors: The Emergence of the Smart Meter

In many ways, the degree of transformation of distribution system sensing technologies surpasses the transmission system development described previously. Traditionally, electrical equipment installed at the customer point was mainly a meter, chief purpose of which was consumer billing [270]. It counted the total number of kilowatt-hours (kWh) consumed and was read once per billing period. This meant that utilities rarely had access to real-time power consumption data at the grid periphery. Instead, real-time data would originate from feeders and substations that were connected to the SCADA network. The remaining "last-mile" of the grid (between these feeders and electricity consumers) was often managed by practical engineering rules based upon feeder data and the feeder's radial topology. These approaches, however, have limited utility in the presence of DG downstream of the last SCADA-monitored feeder [271, 272]. Furthermore, they are equally inapplicable as demand-side resources begin to participate in demand-response programs [271, 272].

The advent of smart meter technology, however, has greatly expanded the capabilities of demand-side metering technology. First, instead of simply measuring aggregate energy consumption, smart meters measure active power consumption as a temporal variable with a sampling rate of up to 1 Hz [273]. Some smart meters also measure power quality as well as voltage and current phase angles [274]. Such measurements naturally produce significant quantities of data which must ultimately be communicated, processed, and stored in new information technology (IT) infrastructure. Nevertheless, the readings from individual smart meters are valuable because they can be used to make advanced analyses for individual meters or aggregated networks [141, 270].

Second, smart sensors, such as smart meters in advanced metering infrastructure (AMI), monitor a bidirectional flow of power and allow for two-way communication between the utility and the consumer [275, 276]. AMI is a system of technologies that measures, saves, and analyzes energy usage from devices such as smart meters using various communication media [46]. AMI meters have embedded controllers, generally including a sensor, a display unit, and a communication component such as a wireless transceiver, and they are generally powered by the electrical feed that they are monitoring [276]. AMI can also incorporate older systems such as automatic meter reading (AMR) and automated meter management (AMM) [46] in their applications. An older AMR system may be capable of remotely collecting power consumption data, remotely relaying power usage, remotely turning a system on or off, and generating bills with different pricing rates [277, 278].

Most utilities have upgraded their investments from AMR to AMI to install two-way communication in a transition to smart technologies with improved demand-side management capabilities [141]. In 2013, the number of two-way AMI meters overtook the number of one-way AMR meters for the first time [279] and by 2016, there were about 46.8 million AMR meters and about 70.8 million AMI smart meters installed by utilities [279, 280]. As eIoT advances to include demand-side management, older technologies need to be upgraded in order to maximize the benefits of eIoT technologies.

3.1.4.2 Network-Enabled Actuators: Distribution Automation

Although distribution automation was initially implemented in the USA (in the 1970s) to increase reliability and resilience in the face of electrical faults [281], eIoT is placing increased demand for automated power quality and real-time network adjustments. Automated feeder switching provides traditional reliability in response to fault identifications, load control and load management [282]. Distribution automation is important not only for resilience with faults, but also as a solution to today's more dynamic loads. Tools such as automated feeder switching must accomplish network-wide reconfigurations for self-healing operations *and* day-to-day operations with increased load variability [283]. Other tools, such as automated voltage regulation and automated power factor correction, increase efficiency and improve power quality [21, 282]. Optimal load balancing through automation results in decreasing power losses, deferring capacity-expansion investment, and improving voltage profiles [21, 283].

Automation in distribution is a step towards a larger, eIoT-enabled smart grid that integrates microgrids for optimal performance [281, 282]. The DOE's Smart Grid Investment Grant (SGIG) Program made advances in distribution automation as an imperative to modernize the electric grid [21]. Partly funded by the American Recovery and Reinvestment Act (ARRA), utilities in the SGIG program installed 82,000 smart devices to 6500 distribution circuits [21]. Figure 3.8 shows the installations of distribution assets from the program.

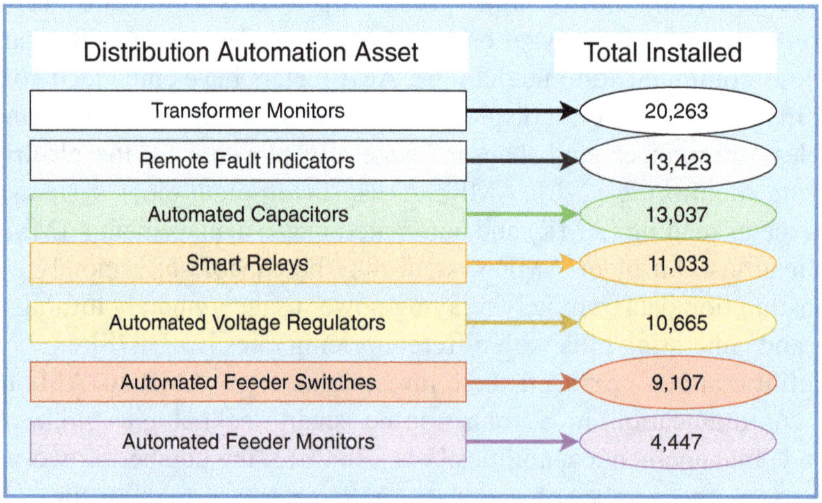

Fig. 3.8 Distribution automation upgrades during the smart grid investment grant program [21]

3.1.5 Sensing and Actuation of Demand-Side Secondary Variables

The sensing and actuation of demand-side secondary variables serves to empower customers to create energy-aware smart homes [284–286], commercial buildings [287, 288], and industrial facilities [289, 290]. In that regard, eIoT developments should be seen as an energy extension to long-standing efforts for automation. Network-enabled sensors again play the key role of providing insights into electricity consumption patterns with potentially device-level granularity. Network-enabled actuators on these devices can then respond to energy-aware decisions that make trade-offs between consumer preferences and energy consumption.

That said, it is important to recognize that secondary variables on the supply and demand sides are fundamentally different. On the electricity supply side, the need for sensing and actuation is entirely motivated by a single purpose: the generation and sale of electricity. On the demand side, secondary variables describe the behaviors of electricity consumers in the residential, commercial, and industrial sectors. The electrical consumption patterns serve a more fundamental purpose of enabling these sectors to carry out their activities *outside* of the electricity sector. Consequently, an effective implementation of eIoT on the demand side always needs to answer the question *"What is the electricity used for?"*. For example, a production facility that uses 1 kW to run a milling machine will not shed that consumption because it directly contributes to production throughput. In contrast, it may shed 1 kW of a back-office because laptop computers can run on their own batteries. Consequently, the remainder of this section breaks the discussion into the various application of eIoT devices.

3.1.5.1 Energy Monitors with Embedded Data Analytics

While device-level sensing granularity of electricity consumption has become a goal of eIoT, in many cases it is not cost feasible. Instead, energy monitors, particularly in home applications, have developed to fill a much needed gap in the eIoT landscape. They are best understood by comparison to smart meters. Smart meters measure aggregate power approximately every minute, and provide data "outward" to the utility. Energy monitors, in contrast, measure a home's or facility's aggregate power consumption every millisecond (1 kHz), and the data is sent "inwards" to the homeowner or facility manager [291]. The operating principle of an energy monitor is illustrated in Fig. 3.9. The aggregate power consumption consists of several device-specific "signatures" that make it possible via data analytics algorithms to recognize when one device is operating versus another. Such a technique is most effective in differentiating high-consuming devices while less so for small devices. The resulting data can be provided to home owners and facility managers for cost-saving decisions. Home energy monitors are currently available at a variety of price

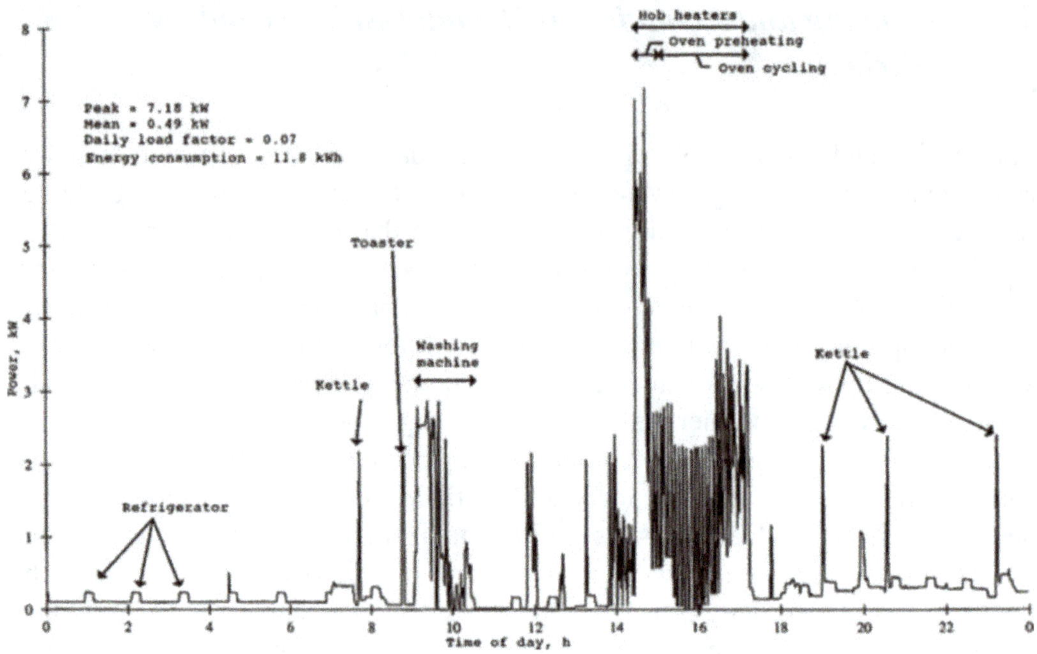

Fig. 3.9 Aggregate profile of household electric power consumption [22]

points from about \$150 to \$400. Continuous gains in energy cost savings outweigh a consumer's initial \$300 investment in a home energy monitoring system.

Meyers, Williams, and Matthews in an article in *Energy and Buildings* [292] used the US Energy Information Administration's Residential Energy Consumption Survey data to estimate the inefficiencies in US home energy usage. The authors estimate that in 2005, 39% of energy delivered to US homes was wasted, costing the homeowners a total of $\$81.5 billion$, or \$733.60 per household on average. Assuming that 41% of the energy inefficiencies could be reduced in part by using a home monitoring system to identify costly consumption behavior, the homeowner could see benefits within the first year of purchasing the system.

3.1.5.2 Network-Enabled Smart Switches, Outlets, and Lights

While energy monitors are relatively effective in resolving an aggregate power consumption profile into its constituent device-level components, they do leave room for further technological development. First, the data analytics algorithms will never resolve devices whose individual power consumption is comparable to the aggregate power consumption's noise level. While this may seem like a trivial issue, in reality, it is important because most facilities have large populations of small devices that together may make up a large part of the total power consumption. Indeed, the Department of Energy has provided practical advice about "phantom loads" that draw electric power simply by remaining idle while plugged in [293].

Phantom loads are costly and inefficient [294, 295]. The average US households waste $100 per year on devices that draw power while not being used [293]. Electronics such as digital video recorders (DVRs) are large users of energy even in standby mode, using 37 W in a home [294]. "Dumb" devices can help decrease phantom loads. For example, connected power strips can make disconnecting groups of appliances easier [294, 296]. Intelligent actuators in home automation overcome inconvenience and human forgetfulness to eliminate phantom loads and provide household savings [297]. Unfortunately, energy monitors do not actuate individual devices without manual intervention. For these reasons, a wide range of smart home devices have developed in recent years to give homeowners device-level visibility and control.

Device-level visibility and control have the potential to transform energy management. eIoT extends to individual home appliances, or production profiles for factories, or HVAC patterns for commercial buildings. The success of such coordination depends on real-time data exchange between smart devices, electricity operations, and the energy consumer [298]. The data includes forecasts of prosumers (dependent on local variables), the energy usage schedule of consumers, and energy-management signals from economic and operation centers [298]. A smart scheduler can then act autonomously to collect data and control devices without active consumer engagement [298]. In so doing, it smooths a household's demand curve and optimizes energy costs [298].

In essence, a smart scheduler is designated as a two-way communication device that synthesizes cost data and appliance profiles to ensure that a household's aggregate consumption does not exceed a predefined limit [298]. The scheduler can shed or defer loads by sending "off," "on", "pause," and "resume" signals to flexible appliances [298]. Hourly profiles can be developed from historical data of the appliances within a month, and it can be determined which appliances are used by a household [298]. Finally, a smart scheduler can act as a load aggregator with the potential to communicate with time-dependent retail and wholesale markets [298].

Perhaps the most common of smart home devices are smart outlets, switches, and lights. Smart outlets are used to cut off phantom loads at the source, without the inconvenience of unplugging appliances. Smart switches can operate by a button, or remotely through apps or a timer [299]. Motion sensors can detect room occupancy and switch lights on and off accordingly [297]. In addition to energy-efficient bulbs (see [300]), there are smart bulbs that can save energy by customizing brightness or color to a set schedule [301]. Although smart home devices are more expensive than their traditional alternatives, their annual energy savings are a counterbalance to the initial investment. Within smart homes, these devices offer not just cost savings but also a level of convenience that many homeowners may wish to have. Because of this, the rationale for adoption is not strictly based upon a return-on-investment (ROI).

In commercial and industrial applications, however, the investment decision is often strictly based upon ROI. Nevertheless, these sectors (as discussed in Sects. 4.4.2 and 4.4.1) often have larger, more energy-intensive equipment that make it easier to rationalize the investment of network-enabled sensors and actuators and their associated energy savings. Given that at least 40% of electricity generation is consumed in commercial and residential buildings, it is important to invest in energy-efficient systems that are also capable of participating in demand response [302].

3.1.5.3 Network-Enabled Heating and Cooling Appliances

While smart outlets, switches, and lights can go a long way to reducing demand-side energy consumption, devices that serve a heating or cooling function are the most energy intensive. Reconsider Fig. 3.9. There are clear power consumption spikes associated with refrigerators, kettles, toasters, heaters, and ovens. Furthermore, air conditioners, alone, account for approximately 6% of US electricity consumption and account for about $49 billion in energy costs.

The appliance marketplace has recognized the potential for developing "smart appliance" versions of these devices. Some appliances have an established market for smart products, while others are just forming. For example, smart refrigerators have a broad offering of features/specifications and efficiency capabilities [301]. Their price depends on the variations in size, doors, cooling features, freezing compartments, displays, efficiency, and power usage.

Smaller devices such as toasters and kettles are emerging as niche tech products. A smart kettle or coffee maker can connect to a smart home hub or to a smart phone app via WiFi, 3G, and 4G to program water temperatures [303, 304]. While the kettle doesn't draw less energy, the scheduling feature has the opportunity to reduce unneeded energy usage. Similarly, a smart toaster can connect to an app on your phone through Bluetooth that enables the remote adjustment of the cooking timer, and return notifications when the toast is ready [305–307]. Smart ovens are another appliance that can connect to smartphone apps to schedule cooking, measure cooking temperatures, and engage either pre-set or customized cooking programs [308]. There also exist smart all-in-one filter, heating, and cooling devices that are able to measure and transmit the temperature and air quality of a room to a mobile app. These values can then be scheduled and controlled in several automated and semi-automated modes [309, 310].

In all these cases, these network-enabled heating and cooling appliances are automated with sensing and software capabilities to optimize their control and performance. Once network-enabled, these devices can be operated remotely to operate at the best possible time regardless of the user's presence. For example, electrified HVAC systems have used a technique called pre-cooling [311]. Instead of cooling a building at the hottest time of the day, the building can be cooled to an artificially low temperature earlier so that it warms but remains at a comfortable temperature during the peak.

Such a technique dramatically reduces electricity consumption because air conditioners are more energy intensive at high ambient temperatures [312]. This technique can be further enhanced with a system that receives and responds to (readily available) weather predictions [311]. Furthermore, smart thermostats can use georeferencing to match the global positioning system (GPS) on a homeowner's phone to the home's thermostat [313]. The device then activates the air-conditioning system based on the phone's proximity and expected time of arrival, and it deactivates the air-conditioning system otherwise.

3.1.5.4 The Electrification Potential of eIoT

Beyond these traditional electrical devices, it is important to recognize the *electrification potential* of eIoT. Figure 1.3 shows a Sankey diagram for the American energy system. Electricity consumption accounts for just $12.6 quads$ of the $97.3 quads$ total. This means that in order to make radical improvements in decarbonization, many of the energy uses that rely directly on fossil fuels must first be electrified so that they will have the potential to be powered by renewable energy sources. In this regard, the transportation sector with $27.9 quads$ of energy consumption (28.7% of the US total) is the first candidate for electrification. Of this quantity, electrified transportation accounts for only $0.03 quads$ (or 0.1% of the transportation total). The manufacturing sectors also consume $24.5 quads$ of energy (25.2% of the US total). Of this quantity, electricity for manufacturing accounts for only $3.19 quads$ (or 13.0% of the industrial total). Finally, the residential sector consumes $11.0 quads$ of energy (11.3% of the US total). Of this quantity, electricity for residential use accounts for only $4.8 quads$ (or 43.6% of the residential total). In all of these cases, a switch from fossil fuels to electricity as an energy source can have a large decarbonization impact [24].

3.1.5.5 Net-Zero Homes: Electrification of Residential Energy Consumption

In residential applications, eIoT can directly support the electrification to achieve homes with net-zero carbon emissions. Returning to Fig. 1.3, the residential consumption of natural gas and petroleum accounts for 5.56 quads of energy, much of which goes to heating applications. Rather than using fossil-fuel furnaces and boilers, net-zero homes [314] often use air [314] and water [314] heat pumps with electricity as an energy supply.

From an energy balance perspective, heat pumps are often twice as efficient as simple resistive electric heating, boilers or furnaces [315]. These energy efficiencies translate directly into significant cost savings as well. Furthermore, recent generations of heat pump technology have embraced IoT [316]. They can be either controlled directly from a smartphone or interfaced with a smart thermostat. Such implementations allow homeowners to tune heating schedules so that they coincide with their home (or even room) occupancy for added savings. The introduction of smart heat pumps also facilitates their usage in active demand-response schemes and their coordination with rooftop solar energy.

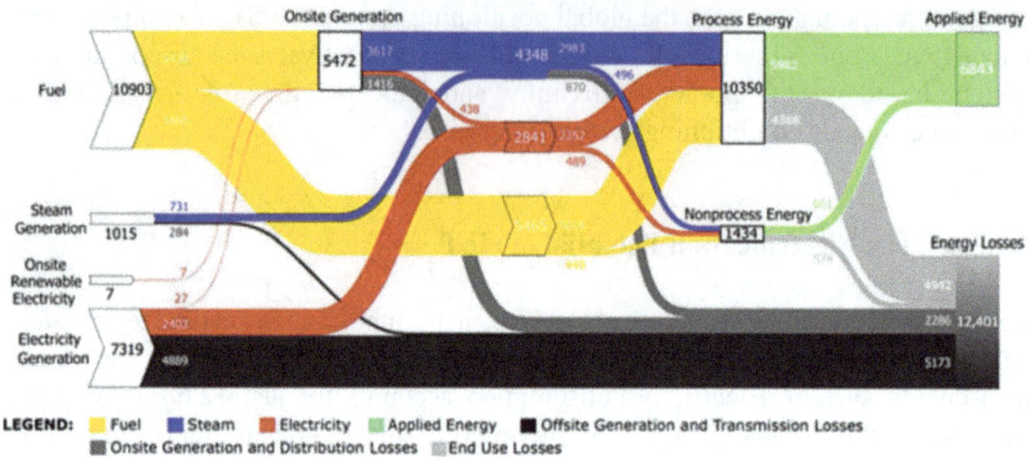

Fig. 3.10 Sankey diagram for the energy consumption (TBtu) of the US manufacturing sector [23]

3.1.5.6 Net-Zero Industry: Electrification of Industrial Energy Consumption

eIoT can have a similar role in the electrification of industrial energy consumption. Unlike residential applications, the electrification of industrial energy usage must (1) strictly follow an ROI rationale and (2) match the required manufacturing processes of the industrial facility. Nevertheless, many industrial sectors have already invested significantly into IoT technologies for supply chain management. Extending these efforts towards energy management is a logical next step.

In 2010, the US Department of Energy conducted a manufacturing energy consumption survey detailing how much of each type of energy was consumed for all major manufacturing sectors [23, 317, 318]. Figure 3.10 shows the associated Sankey diagram for the manufacturing sector in aggregate. It shows a heavy reliance on fossil fuels for steam generation and process heating [23]. In many cases, these fossil-fuel options can be replaced with their electrified alternatives. Figures 3.11 and 3.12 summarize the cost and payback periods of such electrification alternatives for a wide variety of manufacturing sectors. Furthermore, these proposed electrification technologies should be considered as an integral part of eIoT and lend themselves to energy-management practices within the manufacturing plant and the electric grid as a whole [24].

3.1.5.7 Connected, Automated, and Electrified Multi-Modal Transportation

Finally, the transportation sector represents one of the most prominent applications of eIoT. This is due in large part to three fundamental technological shifts that have the potential to transform the sector as a whole [319]: connected automation, electrification, and IoT-based ride sharing.

Subsector	Implementation Cost Range ($000) | Simple Payback Range (Years)							
	Use Immersion Heating in Tanks, Melting Pots, etc.		Convert Liquid Heaters from Underfiring to Immersion or Submersion Heating		Replace Fossil Fuel Equipment with Electrical Equipment		Replace Gas- Fired Absorption Air Conditioners with Electric Units	
	Cost ($x1000)	Payback (Years)	Cost ($x1000)	Payback (Years)	Cost ($x1000)	Payback (Years)	Cost ($x1000)	Payback (Years)
Agriculture	—	—	1	2	—	—	—	—
Food	0.1— 80	<1— 3	0.8— 14	1— 4	0.001— 77	<1— 1	—	—
Textile mills	—	—	—	—	40— 144	1	—	—
Lumber and wood products	—	—	—	—	1— 17	<1— 3	—	—
Paper	—	—	—	—	6— 68	<1— 5	—	—
Printing	0.02	<1	—	—	5	4	—	—
Chemicals	25	2	—	—	2— 31	<1— 5	—	—
Petroleum Refining	—	—	—	—	2— 963	<1— 3	—	—
Rubber and plastics	1— 66	<1— 2	—	—	0.7— 50	<1— 3	3	5
Stone, clay, glass, and concrete	—	—	4	—	2— 6	<1— 1	—	—
Primarfy metals	—	—	—	—	10— 2,000	<1— 6	—	—
Fabricated metal products	1— 55	<1— 2	12	<1	0.8— 150	<1— 5	—	—
Machinery and computer equipment	1	1	2	<1	0.3— 55	<1— 5	—	—
Electronic and other electrical equipment	—	—	0.8— 37	4— 6	0.2— 161	<1— 2	—	—
Transportation equipment	26— 38	2— 5	—	—	0.6— 100	<1— 3	—	—
Miscellaneous manufacturing	—	—	—	—	11— 384	<1	138	3
Recommendation Average	50	2	11	3	84	2	71	4

Fig. 3.11 Summary of manufacturing sector electrification alternatives (adapted from [24])

First, vehicles (of all types) are increasingly outfitted with connectivity solutions so as to become a veritable part of IoT [320–323]. At first vehicle connectivity was simply for emergency roadside assistance and extensions of the driver's mobile phone capabilities [324, 325]. However, the connectivity solutions have greatly expanded in the context of vehicle automation. Adaptive cruise control, where a vehicle's automatic cruise control responds in congested conditions to the fluctuating speed of the car in front, has given rise to a plethora of *vehicle-to-vehicle* connectivity applications [324–327].

Whereas, the first application of adaptive cruise control was driver convenience, it is now being developed for its potential environmental benefits. Research is underway to enable automated vehicle platoons where vehicles automatically follow each other *at short range* so as to reduce overall road congestion and save fuel consumption by aerodynamically drafting. Such automated solutions motivate the need for *vehicle-to-infrastructure* as well. Beyond highway driving, there remains a significant need to reduce traffic congestion, improve air quality, and reduce energy consumption in congested city roads [328, 329].

One important challenge is the coordination of road intersections. Traffic light scheduling, whether it is done statically or dynamically in response to road congestion, has long been an area of extensive research [330–332]. And yet,

Subsector	Implementation Cost Range ($000) | Simple Payback Range (Years)					
	Use Electric Heat in Place of Fossil Fuel Heating System		Replace Hydraulic/ Pneumatic Equipment with Electrical Equipment		Use Heat Pump for Space Conditioning	
	Cost ($x1000)	Payback (Years)	Cost ($x1000)	Payback (Years)	Cost ($x1000)	Payback (Years)
Agriculture	—	—	26— 69	—	—	—
Food	—	—	0.4— 326	<1— 6	4— 22	<1— 5
Textile mills	—	—	2	2	—	—
Lumber and wood products	—	—	0.2— 242	<1— 2	11	—
Paper	3— 80	<1— 2	0.8— 23	<1— 4	—	—
Printing	—	—	4— 19	<1— 2	2— 68	1— 4
Chemicals	0.7— 5	<1	0.9— 32	3— 8	—	—
Petroleum Refining	—	—	1— 80	3— 4	—	—
Rubber and plastics	—	—	0.07— 252	<1— 5	0.7— 1,293	<1— 2
Stone, clay, glass, and concrete	2— 6	<1— 2	1— 18	<1— 2	6	3
Primarfy metals	6— 190	<1— 4	0.2— 110	<1— 3	—	—
Fabricated metal products	2— 146	<1— 3	0.7— 92	<1— 5	1— 429	4— 25
Machinery and computer equipment	0.6— 1	<1— 2	0.2— 27	<1— 3	10— 335	2— 4
Electronic and other electrical equipment	325	3	1— 25	1— 4	7— 152	6
Transportation equipment	2	—	0.9— 14	<1— 1	10— 96	3— 7
Miscellaneous manufacturing	—	—	0.5	<1	182	5
Recommendation Average	**38**	**1**	**24**	**2**	**128**	**5**

Fig. 3.12 Summary of manufacturing sector electrification alternatives (adapted from [24])

solutions like traffic lights retain a *driver-in-the-loop* control paradigm. More recent research envisions the elimination of traffic lights so that the intersection itself can coordinate the crossing of vehicles and potentially even pedestrians [333–336]. Vehicle automation has been classified into five levels of technology development with some analysts predicting full Level 5 automation by 2030 [337–340].

It is important to recognize that these developments toward connected automation exist in all modes of transport. Planes and trains have been automated to varying degrees for decades [46, 341–343], while buses and trucks are directly benefiting from developments in the car market [344]. Nevertheless, the shift toward connected and automated road vehicles is important because of its share of overall vehicle miles traveled [340] and because of the difficulty of its coordination and control problems.

As a second fundamental shift in technology, electrified transportation greatly complements the benefits of connected and automated vehicles. As mentioned, in Chap. 1, the electrification of transportation is one of the five identified energy-management change drivers. Electrified transportation supports energy consumption and CO_2 emissions reduction targets [41, 345–348]. Relative to their internal combustion vehicle counterparts, EVs, whether they are trains, buses, or cars, have a greater "well-to-wheel" energy efficiency [348, 349]. They also have the added benefit of not emitting any carbon dioxide in operation and rather shift their emissions to the existing local fleet of power generation technology [42]. Furthermore, the technical, economic [350–352], and social barriers [82, 353] to their adoption have eased. Despite continuing challenges in battery technology [354–356], a wide variety of battery chemistry options have emerged leading to greater capacity and subsequently vehicle ranges [357–359]. Fast chargers have also been introduced into the market which allow 80% of the battery capacity to be charged in 30 min [360–362]. From an economic perspective, both plug-in hybrid EVs and battery-EVs show significant learning rates and cost improvements over time [73, 352]. There also exist significant improvements in public attitudes [363–366] and social transition rates [82, 349, 353, 367]. As a result, a number of optimistic market penetration and development studies have emerged for a wide variety of geographies [368–374]. Consequently, supportive policy options have taken root worldwide [363, 375, 376].

The true success of electrified (multi-modal) vehicles depends on its successful integration with the infrastructure systems that support them. From a transportation perspective, plug-in electric cars may have only a short range of 150km [365], but it may still require several hours to charge them [377]. This affects when a vehicle can begin its journey and the route it intends to take. From an electricity perspective, the charging loads can draw large power amounts that may exceed transformer ratings, cause undesirable line congestion, or cause voltage deviations [378–381]. These loads may be further exacerbated temporally by similar charging patterns driven by similar work and travel lifestyles or geographically by the relative sparsity of charging infrastructure in high-demand areas [380]. This *transportation-electricity nexus* (TEN) [31, 89–91, 382] requires new assessment models whose scope includes the functionality of both systems. Recent works have also proposed axiomatic design as a means to model large systems such as the transportation and manufacturing systems [383–387]. As the complexity of these systems increases, it becomes more relevant to consider their resilience while especially focusing on flexibility and reconfigurability [382].

Relatively few studies have considered this coupling from an operations management perspective. A simplified study based on the city of Berlin has been implemented on the multi-agent transport simulation (MATSIM) [362]. Meanwhile, the first full-scale study was completed in the city of Abu Dhabi [379, 388–390] using the clean mobility simulator [391]. A third study focused on the differences between conventional plug-in and online (wireless) EVs [31]. More recently, a performance assessment methodology for multi-modal electrified transportation has been developed that integrates the methodologies of previous studies [91]. An older review compares a variety of open source transportation modeling tools [392].

IoT-based ride sharing, as the third fundamental shift in transportation technology, has the potential to dramatically intertwine vehicle automation and electrification. It expands the transportation options available to travelers so that even incumbent paradigms of vehicle ownership are questioned [393–395]. Travelers, particularly in large cities, are now more likely to rely on a combination of transportation modes to arrive to their destination. In some cities, IoT-based ride sharing has already shifted transportation behavior from the traditional use of private cars [393, 395]. This work, however, argues that IoT-based ride sharing is likely to converge with eIoT-based energy management because their underlying decisions are fundamentally coupled.

Consider an EV rideshare fleet operator [379, 388–390]. They must dispatch their vehicles like any other conventional fleet operator, but with the added constraint that the vehicles are available after the required charging time. Once en route, these vehicles must choose a route subject to the nearby online (wireless) and conventional (plug-in) charging facilities. In real-time, however, much like gas stations, these charging facilities may have a wait time as customers line up to charge. Instead, the EV rideshare driver may opt to charge elsewhere. Once a set of EV rideshare vehicles arrive to a conventional charging station, the EV rideshare fleet operator may wish to implement a coordinated charging scheme [45, 80, 81, 396–404] to limit the charging loads on the electrical grid. The local electric utility may even incentivize this EV rideshare operator to implement a "vehicle-to-grid" scheme [82, 362, 405] to stabilize variability in grid conditions.

These five transportation-electricity nexus operations management decisions are summarized in Table 3.1 [31, 89]. The integration of such decisions in a coordinated fashion ultimately forms an intelligent transportation-energy system (ITES) [389]. Naturally, significant research remains on how to best integrate these decisions so that they achieve operational benefits in both the transportation and electric power systems. More recently, studies have focused on the design of smart cities and their core infrastructures such as transportation, district heating and cooling (DHC), and electric power grid. Specifically, hetero-functional graph theory has been introduced as a more advanced means of studying coupled infrastructures such as the TEN [406, 407].

Table 3.1 Intelligent transportation-energy system operations decisions in the transportation-electricity nexus [31]

- **Vehicle dispatch**: When a given EV should undertake a trip (from origin to destination)
- **Route choice**: Which set of roads and intersections it should take along the way
- **Charging station queue management**: When and where it should charge in light of real-time development of queues
- **Coordinated charging**: At a given charging station, when the EVs should charge to meet customer departure times and power grid constraints
- **Vehicle-2-grid stabilization**: Given the dynamics of the power grid, how can the EVs be used as energy storage for stabilization

3.1.6 Network-Enabled Physical Devices: Conclusion

This section has provided an extensive discussion of the state of the art in network-enabled physical devices, whether they are network-enabled sensors or actuators in the control loop. In order to organize the discussion, Fig. 3.2 was used to distinguish between primary and secondary electric power system variables. In all, four major categories of network-enabled devices were discussed.

- Section 3.1.2 addressed the (traditional) primary variables in the transmission system.
- Section 3.1.3 discussed the concerns around the secondary variables associated with wind, solar, and natural gas generation.
- Section 3.1.4 returned to the primary variables in the distribution to address smart meters and other "grid modernization" technologies.
- Section 3.1.5 discussed smart homes, industry, and transportation in the context of demand-side secondary variables.

3.2 Communication Networks

3.2.1 Overview

The tremendous heterogeneity of network-enabled devices described in the previous section demands advancements in communication networks to route sensed information to control and decision-making entities. Because these devices vary greatly in size, power consumption, use case, and on-board computing, new types of networks will emerge that can enable two-way flows of information. Consequently, these networks must have different scope and ownership.

Figure 3.13 shows several network areas relevant to the electric power system. Starting at the center of the grid, utility networks are the communication backbone for grid operations. Wide-area networks (WAN), as the largest in geographical scope, encompass centralized generation, transmission, and substations under the utility's domain. Moving "downstream" from the substations, neighborhood area networks (NAN) are of intermediate scope and use public and commercial telecommunication networks throughout the distribution network. The NAN serves AMI, meter aggregations, DER, and microgrids, which can also include utility participation. Finally, local area networks (LAN) address the private communication scope of residential, commercial, and industrial entities. These networks can encompass subnetworks that connect to a NAN or directly to the public internet [25]. The following definitions apply to the rest of this discussion:

Definition 3.3 (Commercial Telecommunication Network) A telecommunication network that is owned and operated by a commercial telecommunication company. ■

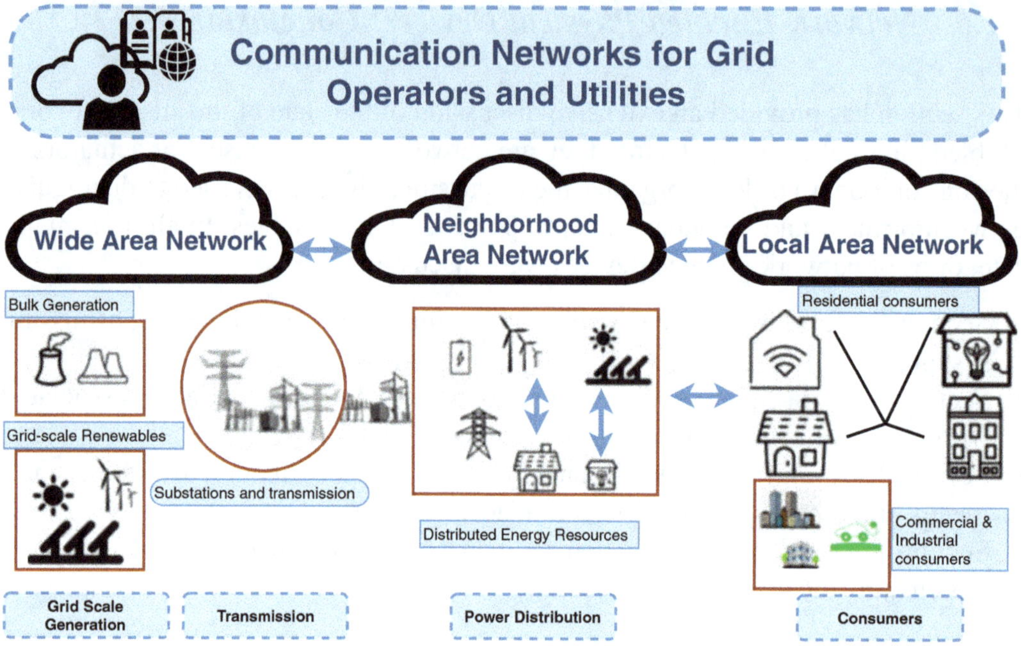

Fig. 3.13 LAN, NAN and WAN networks across the electric power system (adapted from [25])

Definition 3.4 (Private Network) A network that is owned and operated by a private entity, be it residential, commercial, or industrial. In scope, a private network may be a WAN, NAN, or LAN. It may use interoperable, standard, or proprietary technologies. ■

Definition 3.5 (Proprietary Network) A network that is not based upon an interoperable standard. Note that some networks may use open standards but are not interoperable because the standards themselves are not interoperable. ■

The development of mature eIoT communications is likely to be a gradual migration process. Traditionally, the power system has used private networks within the jurisdiction of grid operators and utilities. These include transmitted data over wired networks (e.g., power-line carrier and fiber optics) as well as wide-area wireless networks such as SCADA (supervisory control and data acquisition). However, with "grid modernization," commercial telecommunication networks are increasingly playing a role.

Cellular data networks, and in particular 4G and long-term evolution (LTE), have the potential to transmit relatively high bandwidth data across long distances. Furthermore, WiMax networks can provide connectivity at the grid periphery at the neighborhood length scale. Finally, a large part of eIoT will require local area networks, be they wired Ethernet, WiFi, Z-wave, ZigBee, or Bluetooth. Naturally, industrial energy-management applications continue to leverage preexisting industrial network infrastructure in addition to these local area network options. Technological developments in communication networks are most likely to occur as a gradual migration rather than a swift shift from one technology to another. Furthermore, these developments are likely to occur in parallel so as to become complementary and mutually co-existing.

- Tables 3.2, 3.3, and 3.4 summarize the eIoT communication networks discussed in this section.
- Section 3.2.2 discusses grid operator and utility networks.
- Section 3.2.3 discusses telecommunication networks.
- Section 3.2.4 discusses local area networks.

3.2.2 Grid Operator and Utility Networks

Grid operator and utility networks use a range of legacy communication systems and technologies that are very much a product of the regulated electric power industry from several decades ago [428]. Nevertheless, technological developments in data acquisition, data analysis, and renewable energy generation are now pressuring grid communication systems to evolve and adapt. For example, the variability of renewable energy generation (discussed in Chap. 2) requires automatic control whose data rates are faster than what legacy communications systems are able to provide. This section highlights some of these traditional technologies so as to contextualize the discussion of eIoT communication technologies.

This section categorizes grid operator and utility communication into wired and wireless networks, each with their respective trade-offs and applicability within the electric system.

- For wired communications, power-line carrier networks and fiber optics are covered in Sect. 3.2.2.1 [412]. Wired communications are relatively reliable and secure and very much represent the historical default for electrical utilities. However, their widespread deployment is associated with high rental fees and installation costs [106, 412]. Grid operators and utilities have also made extensive use of wireless networks, which in comparison have lower cost and reliability. Their flexibility and ease of installation, however, often supports their adoption.
- Section 3.2.2.2 is devoted to SCADA-based wide-area monitoring systems as a traditional wireless power grid communication network.
- Section 3.2.2.3 then delves into the emerging world of low power wide-area networks (LPWAN).
- Section 3.2.2.4 discusses the wireless smart utility network (Wi-SUN) as a new development. Other types of wired and wireless communication networks are discussed more deeply in the context of commercial telecommunication and local area networks.

Table 3.2 Communication networks for grid operators and utilities

Grid operator and utility networks								
Network	Application	Data rate Data rate	Distance Distance	Wired/ wireless	Standard	Topology	Advantages	Disadvantages
PLC	Transmission, distribution [409]	10 kbps–200 Mbps [409]	200–3000 m [410]	Wired	(1) HomePlug [408] (2) Narrowband [408] (3) IEEE P.1901 [408] (4) IEEE 1901 [408] (5) HomePlug AV [408] (6) High definition power-line communication (HDPLC) [408] (7) ITU-T G.9960 standard [408] (8) CENELEC EN 50065 standard [408]	Star	(1) Wide coverage [408] (2) Low cost [408] (3) Flexibility and range [408] (4) Mobility [408] (5) Easy installation [408] (6) Stability [408] (7) Located where the circuits are required [411] (8) Equipment installed in utility owned land, or structures [411] (9) Economically attractive for low numbers of channels extending over long distances [411]	(1) High noise over power lines [408] (2) Capacity [408] (3) Open circuit problem [408] (4) Attenuation and distortion of signal [408] (5) Inadequate regulations for broadband PLC [408] (6) Not interoperable [408] (7) Not independent of the power distribution system [411] (8) Carrier frequencies often not protected on a primary basis [411] (9) Expensive on a per-channel basis compared to microwave [411] (10) Will not propagate over open disconnects [411] (11) Inherently few channels available [411]

(continued)

Table 3.2 (continued)

Grid operator and utility networks								
Network	Application	Data rate	Distance	Wired/ wireless	Standard	Topology	Advantages	Disadvantages
Fiber optics	Transmission, distribution [412]	155 Mbps–40 Gbps [410]	60,000–100,000 m [410]	Wired	PON [410] WDM [410] SONET/SDH [410]	Star	(1) Reliability [410] (2) High quality [410] (3) Immune to electromagnetic interference [411] (4) Immune to ground potential rise [411] (5) Low operating cost [411] (6) High channel capacity [411] (7) No licensing required [411]	(1) Different skill set for fiber optics needed [411] (2) Expensive test equipment [411] (3) Inflexible network configuration [411] (4) Cable subject to breakage and water ingress [411]
Digital subscriber line (DSL)	Transmission	1–100 Mbps [410]	1500–5000 m [410]	Wired	ADSL [410] HDSL [410]	Star	(1) Low investment cost [106] (2) High speed [106] (3) High bandwidth [106]	(1) Non-ownership of infrastructure can cause reliability issues [106] (2) May not be available in remote locations [106]
LoRaWAN	Distribution	0.3–27 kbps [417]	2–5 km 15 km [417]	Wireless [415, 416]	AES-128	Star	Low power [415, 416] Fairly high data rates [417] Low cost [417] Flexible and open network [417]	No guaranteed message receipt
SigFox	Transmission, distribution, transmission	100 bps [417]	up to 50 km [419]	Wireless [415, 416]	BPSK [418, 419]	Star	Low power Bidirectional communication [418, 419] Low cost	Network owned by SigFox [417] Only 14 packets/day per device [417] Only 12 bytes per transfer [417] Signal frequency prone to interference [420]

SCADA-Wide area monitor-ing	Transmission	9.6–115.2 kbps [413]		Wired and wireless	Modbus [413] DNP3 [413] IEC 61850, [138]	Star/Peer-to-peer	(1) Minimal latency for WAMs [414] (2) PMUs provide continuous measurements [414] (3) Faster data rate collection than traditional SCADA [414]	(1) Low bandwidth [272] (2) Security concerns with open communication protocols and unintentional connection with other networks [413] (3) SCADA devices have limited computational abilities [272]
NB-IoT	Transmission, distribution	<100 kbps [422]	<10 km	Wireless	3GPP R13 [421] LTE [421, 422] GSM [421, 422]	Star	Wide area Bidirectional [421, 422] Low power consumption [421, 422] Massive connections, 50k [421, 422]	Relatively expensive [421, 422]
Ingenu	Distribution	624 kbps 156 kbps [417]	5–6 km [417]	Wireless	RPMA [417, 418]	Star	Higher data rates [417]	Lower range [417] Higher power consumption [417]

Table 3.3 Telecommunication networks

Telecommunication networks								
Network	Application	Data rate	Distance	Wired/ wireless	Standard	Topology	Advantages	Disadvantages
Cellular Data	Distribution (NAN) [423]	14.4 kbps–100 Mbps [410]	50,000 m [410]	Wireless	GSM, 2.5G, 3G, 4G, LTE [410]	Meshed	(1) LTE is characterized by high (2) Reliability and low latency [424] (3) Scalability [424] (4) LTE can serve as the default or backup network [424]	(1) Cellular service providers face challenges from a growing mobile, user base which may effect all users [425] (2) Future uses may need faster data rates than 4G networks can provide [425] (3) Poses increased security threat by being a public network [423] (4) Sharing the network may result in decreased performance [106]
Wi-Max	Distribution [138]	75 Mbps [410]	50,000 m [410]	Wireless	802.16 [410]	Meshed	(1) Control of the proprietary network [424] (2) Bandwidth and range suited for NAN [412, 424] (3) Relatively high data rates [424] (4) Low latency [424] (5) Relatively low deployment and operating costs [424] (6) Can support real-time data transfers needed for smart meters [424]	(1) Initial infrastructure cost for radio equipment [424] (2) Radio equipment requires optimizing the number of station installations and quality of service requirements [424]

Table 3.4 Local area networks

Private area networks								
Ethernet	Home and building automation	10 Mbps–10 Gbps [410]	100 m [410]	Wired	802.3× [410]	Star	(1) High data rate [410] (2) Range of data rates depends on cable used [426]	(1) Inflexibility of topology [426] (2) Unlikely to find connections on home appliances [426] (3) High cost [426] (4) High power requirements [426]
Wi-Fi	Home and building automation	2–600 Mbps, [410]	100 m [410]	Wireless	802.11× [410]	Meshed	(1) High speed [427] (2) Used for a variety of personal devices (interoperability) [426] (3) High bandwidth [426]	(1) Not meant for moving devices, and though not intended for metropolitan areas it has been extended to larger areas [427] (2) High energy consumption [423]
Z-Wave	Home automation	40 kbps [410]	30 m [410]	Wireless	Z-Wave	Meshed	(1) Low cost [426] (2) Low power consumption [426]	(1) Low bandwidth [426]
ZigBee	Home and building automation	250 kbps [410]	100–1600 m	Wireless [410]	Zigbee, Zigbee Pro	Meshed [410]	(1) Long range in HAN [427] (2) Low power consumption [412], rates [412, 423] (3) Long range in HAN [412]	(1) Devices have limited internal memory, limited processing capability, and low data (2) Weak security [412]
Bluetooth	Home automation	721 kbps [410]	100 m [410]	Wireless	802.15.1 [410]	Meshed	(1) Low power consumption [412]	(1) Weak security [412]

3.2.2.1 Wired Communications: Power-Line Carriers and Fiber Optics

Grid operators and utilities have used power-line carriers and fiber optic cables in transmission and neighborhood distribution applications. Over numerous decades, these technologies have undergone several upgrades from their original implementations, including from analog to digital communication [411]. In the past, the primary need for wired communication was fairly limited to application such as timely and efficient fault detection. This meant that communication systems needed to adhere to stringent cost rationales. A common strategy was to make use of existing utility-owned power poles or rent telecommunication poles to route information back to a control center [411]. This required wired communication systems often to match the radial topology of the underlying physical infrastructure.

Power-line carrier (PLC) communication uses power cables as a medium for data signal transmission [412]. It falls into four categories:

- Ultra-narrow band power-line communication (UNB-PLC)
- Narrowband power-line communication (NB-PLC)
- Quasi-band power-line communication (QB-PLC)
- Broadband power-line communication (BB-PLC)

Depending on PLC technology, data transfer speeds range from 100 Bps to 1.8 Gbps [409, 423]. The X-10 PLC protocol was influential in establishing narrowband PLC communication in the USA [409]. Since then, today's NB-PLC standards include PoweRline Intelligent Metering Evolution (PRIME) (ITU-T G.9904), G3-PLC (ITU-T G.9903), IEEE 1901.2 2013, and ITU-T G.hnem [409]. The 63-PLC smart-grid applications have a 1.3–8 km range [409]. Depending on modulation type, this PLC could have a bandwidth of 30–35 kilobits per second (kbps) or 100 kbps [409]. PLC technologies are used in a diverse array of applications including home, transmission, and connective energy systems [409, 429]. For example, the G3-PLC standard has been used experimentally in the mid-voltage range with several topologies [429]. It has also been used to enable "smart grid" technologies such as AMI, vehicle-to-grid communications, demand-side management, and remote fault detection [408]. Broadband PLC, in particular, is suitable for local area networks (LANs) and AMI applications in the smart grid because it has higher bandwidth (but shorter range) as compared to narrowband PLC [409, 423].

In recent years, utilities have applied optical fiber communication as an upgrade to aging infrastructure [412]. Optical fiber is mainly used as a "backbone" distribution communications network, in what is called fiber-to-pole networks [412]. Optical fiber is characterized by high transfer rates, good stability, strong anti-interference ability, flexible network configuration, large-system capacity, and high reliability [412]. The data rate of optical fiber ranges from 155 megabits per second (Mbps) to 40 Gbps [410]. However, its implementation is a large investment because it requires relatively expensive testing and highly skilled installation and maintenance [411, 412].

The wide-area deployment of wired technologies (that is, PLC and optical fiber) is costly but does provide the benefits of communications capacity, reliability, and security [412]. Some utilities have also installed specialized communication networks according to their specific technical and economic needs. Such specialized lines are mainly composed of twisted-pair cable and provide for small capacity, high reliability, low transfer rate, and moderate anti-interference for a small investment [412].

3.2.2.2 SCADA Networks and Wide-Area Monitoring Systems

SCADA was developed in the 1950s because utilities needed a way to gather power output data from the scattered geography of the electric grid's sensing endpoints to conduct load-frequency control and economic dispatch [101]. SCADA systems now communicate commands and system state data back and forth between utility control stations and individual substations within several seconds [428]. Due to the expansive geographical area covered by the transmission system, monitoring is a large task, and has special sensor communication requirements. SCADA systems have increased "openness" by connecting to wide-area monitoring systems (WAMS) and other networks through proprietary connections and the Internet [430]. This point is emphasized since connection to the internet is an important stepping stone in the development of eIoT.

The SCADA system in actuality uses a combination of wired and wireless technologies. Wired options include telephone lines and optical fiber; wireless alternatives include microwave and ultra-high frequency (UHF) radio [19]. The choice of implemented technology depends on an individual system's needs for data rate, cost, and data security [19]. With traditional technologies, the data rate is typically 9.6–115.2 kbps [413]. SCADA protocols are based on IEEE C37.1 for the communication between remote terminal unit (RTU) and the master terminal unit (MTU) [19]. Traditionally, SCADA allows for serial communication between master and remote terminal units, but newer hybrid protocols allow peer-to-peer communication [272, 413]. These protocols include Modbus, DNP3, PROFIBUS (from standards IEEE 11674, IEEE 61158), DeviceNet, ControlNet, and Fieldbus [272].

The advantages and disadvantages of operating a legacy SCADA system are typical of any aging communication technology. On the one hand, the operating costs are small relative to the initial investment in infrastructure. On the other, the bandwidth and computational capability is relatively low [272]. Furthermore, as SCADA networks have developed, they have suffered unintentional negative consequences. Since the 1990s, utilities began transitioning from closed proprietary networks to interconnected and open internet-based networks [430]. The push for open communication protocols has increased network accessibility and consequently the potential for connection to other networks [413]. This is also an effect of custom networks being standardized so as to be sold as off-the-shelf SCADA systems [430]. As proprietary networks are turned into open networks, and peer-to-peer communication among SCADA devices increases, cybersecurity concerns have naturally increased [413].

In addition to SCADA, WAMS are being deployed as a form of complementary *sensor network*. A WAMS is a collection of hundreds of phasor measurement units (PMUs) at various locations in the electrical grid [414]. PMUs have faster data collection rates than SCADA systems, with 30–60 data points per second as compared to SCADA's 1 data point per 1–2 s [431]. Data communications specifications are provided by the IEEE C37.118-2005 standard [414]. A phasor data concentrator (PDC) aggregates measurements from local PMUs through a local communication network, and then routes the data to a utility's core network using proprietary networks [414]. Data transfers from the PMU to the PDC are required to have minimal latency for an efficient smart grid [414]. PMU data are produced continuously and synchronously and are therefore delay-sensitive [414]. Consequently, it must be intelligently scheduled to manage communication load and maintain quality requirements [414].

3.2.2.3 LPWAN Commercial Wireless IoT Technologies

Due to power constraints on remote IoT sensors and actuators, IoT devices need to operate in an energy efficient manner. Recently, commercial applications to support wide-area communication have emerged. Low power wide-area networks (LPWAN) is an umbrella term that encompasses technologies and protocols that support wide-area (> 2 km) communication and consume low power over long periods of time [432]. Data ranges for these devices are from 10 bps to a few kbps [433]. LPWAN networks must meet the following considerations [433]. Devices should have the following characteristics:

- Be cheap to deploy
- Operate on very low power
- Function when required, preferably in star topologies
- Ensure secured data transfer
- Have robust modulation.

LPWAN networks will generally include devices, a network infrastructure, protocols, controllers, network and application servers, and a user interface [433]. This service can be provided as a single package or through coordination among multiple providers [433].

LoRa, short for long range, is a physical-layer LPWAN application by SemTech Corporation [434]. The system works in the 902–928 megahertz (MHz) frequency band in the USA and in the 863–870 MHz in Europe [418]. The LoRa system is composed of the PHY layer which is proprietary while the LoRaWAN protocol is an open standard that is managed by the LoRa Alliance which has over 300 members [415, 418, 433]. LoRa chips can be produced by various silicon providers to avoid a single source [433]. LoRa networks follow a star topology to relay messages between end-devices and a central network node [415, 416, 418]. Long-range wide-area network (LoRaWAN) radios are used with low power devices to support low bandwidth and infrequent (128 s) communication over wide areas [415, 416, 432].

This drives down the cost and extends the battery life of the devices. LoRaWAN devices draw no more than 2 μA while resting and 12 mA when listening [415, 416]. LoRaWAN can use a bandwidth of 125 kHz, 250 kHz, or 500 kHz depending on the region, application, or frequency [435]. The data rates can also be determined based on the frequency chosen [435]. These data rates typically range from 0.3 to 27 kbps [417]. It uses the AES-128 algorithm that is similar to the IEEE 802.15.4 standard [435]. LoRaWAN offers two security layers, one for the network layer and one for the application layer [433]. It offers a range of 2–5 km in cities and up to 15 km in suburban areas [417]. Another LPWAN technology is the Symphony Link by Link Labs that is a proprietary MAC layer built on top of the LoRa physical layer. This technology adds vital connectivity to LoRaWAN such as guaranteed message receipt [436]. Applications using LoRa technology in the power industry include radiation leak detection from nuclear power plants [437] and air pollution monitoring for thermal power plant systems [438].

The NB-IoT is narrowband communication system by the Third Generation Partnership Project (3GPP) standards body that was launched in 2016 [439]. It is used for low power, infrequent (over 600 s) communication devices [415, 439]. It supports a star topology [415, 439]. It can operate either in the GSM spectrum or LTE [415, 439]. NB-IoT can be deployed in three operation modes: (1) stand-alone using GSM, (2) in-band where it operates within a bandwidth of a wide-band LTE carrier, and (3) with the guard-band of an existing LTE carrier [439]. Since NB-IoT is based on LTE, hardware reuse and spectrum sharing is possible without coexistence issues [439]. NB-IoT is expected to ensure long battery life (up to 10 years) and to support over 52k low-throughput devices [439]. NB-IoT can cover a range of $<$25 km and offers high accuracy rates [422]. The expected latency for this system is $<$10 s for 99% of the devices [439]. NB-IoT systems are used in applications such as smart metering (gas, water, and electricity), smart parking, smart street lighting, and pet tracking [440, 441]. The NB-IoT forum comprises of over 500 members, contributors, and developers [441].

SigFox was launched in 2009 by the French company SigFox as the first LPWAN application for IoT. Compared to LoRa, SigFox is not nearly as widely used in the USA because its frequency band (900 MHz) is very prone to interference and its transmission time ($\approx$3 s) is greater than the maximum transmission time of 0.4 s that is allowed by the Federal Communications Commission (FCC) [420]. The SigFox physical layer uses an ultra-narrowband technology that uses standard ratio transmission method called binary phase-shift keying (BPSK) going up and frequency-shift keying coming down [418, 419]. The SigFox technology is suitable for applications that require small and infrequent transmission [419]. The first releases were unidirectional but recent versions support bidirectional communication [418, 419]. SigFox offers data rates of 100 bps in the uplink with a maximum payload of 12 bytes [417]. It claims to support about a million connected objects with a coverage range of up to 50 km [419]. SigFox has not been as widely adopted, especially in the USA, due to its limiting transmission characteristics such as a restriction on the number of packets transferred by a device to only 14/day [417].

In the electricity and utility industry, SigFox is used to monitor back-up power supply systems and smart metering (gas, electricity, and water) and for electric pole surveillance [442].

Lastly, Ingenu, formally known as On-Ramp Wireless, works in the 2.4 GHz frequency and has a robust physical layer that allows it to still operate over wide areas [418]. It offers higher data rates compared to LoRa and SigFox [417]. Specifically, it can transmit up to 624 kbps in the uplink and 156 kbps in the downlink [417]. Its coverage is, however, shorter (around 5–6 km) and consumes much higher energy [417]. Ingenu is based on the random phase multiple access (RPMA) [417, 418].

3.2.2.4 Wireless Smart Utility Network

The wireless smart utility (ubiquitous) network (Wi-SUN) is a mesh topology network supported by the Wi-SUN Alliance. The Wi-SUN Alliance was founded in 2012 and comprises of 130 members who include product and silicon vendors, software companies, utilities, government institutions and universities [443]. The goal of the Wi-SUN Alliance is to promote open industry standards for wireless communication networks for both field area networks (FAN) and local area networks (LAN) [443, 444]. It also defines specifications for testing and certifying of said networks to enable multi-vendor interoperable solutions [443]. The Wi-SUN network was developed according to the IEEE 802.15.4g standard that defines physical layer (PHY) and medium access control (MAC) layer specifications [445], TCP/IP and related standards protocols.

Applications for the utility include the provision of field area networks (FANs) for smart metering infrastructures, distribution automation, and home energy management. The Wi-SUN coverage range is 2–3 km making it suitable for NANs [446]. AMI systems can use Wi-SUN technology for multiple meters [446]. Wi-SUN networks are usually laid out in a mesh topology although they support both star and star-mesh hybrid topologies [415]. This allows for enough redundancy in the network to limit single points of failure [415]. This network is deployed on both powered or battery-operated devices [415]. Devices that support mesh networks transmit over a short range and are suitable for applications that require distributed computing. The Wi-SUN mesh networks are self-forming. That is, whenever a new device is added, it immediately finds peers to communicate with and whenever a device disconnects the other devices in the peer-network reroute accordingly [415]. The short-range feature allows for faster and consistent data rates. Wi-SUN devices can perform frequent (up to 10 s) and low-latency communication, and draw less than 2 μA in resting and 8 mA when transmitting [415].

3.2.2.5 eIoT Perspectives on Grid Operator and Utility Networks

Grid operators and utilities have long made use of communication networks to gain situational awareness as an integral part of power systems operations and control. In many ways, the communication technologies described above were deployed as part of a regulated electric power industry. eIoT, however, as has been discussed at length will fundamentally change the nature of power system operations so as to need far more advanced communication system technologies. With the above interoperable LPWAN and Wi-SUN technologies, eIoT communication technologies for grid operators and utilities are likely to improve significantly. Open, interoperable standards also create room for innovation within this area.

One main need is the communication beyond the purview of just the grid operators and utilities. In that regard, communication over power-line carriers, proprietary fiber optics, and SCADA leave many new parties out of the evolving and highly flexible eIoT "cloud" [428]. As the next subsections will discuss, there is much room for these utility networks to be complemented by commercial telecommunication networks and LANs [160, 431]. Such a hybrid communication system architecture is much more likely to meet the new and unprecedented requirements for data access and transfer [447]. Naturally, a shift toward hybrid communication systems brings about very legitimate questions of jurisdiction, ownership, and authority over the data, servers, and communication channels that constitute the system. While it is clear that standards will continue to play a central role in the design of communication systems, it remains unclear what role regulation and legislation will have in these areas. These are still open questions as the grid transforms itself towards an eIoT paradigm.

3.2.3 Commercial Telecommunication Networks

One important trend in the development of eIoT communications is the shift towards commercial telecommunication networks as a complement to existing and dedicated grid operator and utility networks. In many ways, this has been a long-standing trend. The preceding section mentioned that utilities and grid operators have often rented telecommunication poles for wired communications over power-line carriers. A logical technological next step is to switch from power-line carriers to digital subscriber lines (DSL) over the (wired) telephone lines themselves [106]. DSL has high speeds of 1–100 Mbps depending on its type, that is, asymmetric digital subscriber line (ADSL), very-high-bit-rate digital subscriber line (VDSL), and high-bit-rate digital subscriber lines (HDSL) [410].

Although DSL technology is often chosen for smart grid projects because the use of existing telephone infrastructure reduces installation costs [106], the lack of standardization and differing ownership of equipment can cause potential reliability issues related to maintenance and repair [106, 412]. Furthermore, the expansion of telephone infrastructure needs to be cost rationalized in remote applications [106, 412].

Beyond wired telephone lines, eIoT communications is now making extensive use of *wireless* telecommunications networks for essential "smart grid" applications such as AMI-to-utility control center communications [106]. Wireless solutions have relatively very low cost [412] and are easier to implement in less accessible regions [106]. Despite these benefits, wireless options present several challenges including constrained bandwidth, security concerns, power limitations, signal attenuation, and signal interference [106].

With these trade-offs in mind, it is useful to acknowledge the needs of the utilities in choosing the most suitable network. Utility evaluation of communication networks usually involves consideration of the following [412]:

1. Bandwidth
2. Data rates
3. Coverage
4. Reliability of end-to-end connection solutions
5. Associated protocols
6. Integration of existing systems
7. Ease of deployment
8. Management tools
9. Life cycle costs

Section 3.2.3.1 highlights some of the technological developments in cellular data networks, and Sect. 3.2.3.2 covers WiMax networks before discussing their implications on eIoT in Sect. 3.2.3.3.

3.2.3.1 Cellular Data Networks: 2.5G-GPRS, 3G-GSM, 4G, and LTE

Cellular communication systems have provided coverage for data transmission for several decades [157]. They enable utilities to deploy smart metering in a wide-area environment and are a relatively quick and inexpensive option for meter-to-utility as well as distant node-to-node communication [106, 157]. Existing telecommunications infrastructure reduces investment cost and the additional time needed to build communications for a power systems purpose [106]. Systems, such as 2.5G, GSM, 3G, and 4G, are radio networks that communicate via at least one base station transceiver (or cell) per land area [157].

2.5G, also known as general packet radio service (GPRS), is a packet data bearer service over the global system for mobiles (GSM) [427]. User data packets are transferred between mobile stations and external IP networks so that IP-based applications can run on a GSM network [427]. Data speeds can range from 9.6 to 115 kbps by amalgamating unused time slots in the GSM network [427].

The next generation cellular network, 3G-GSM, provides data rates of 144 kb/s to over 3 MB/s [412]. GSM itself is widely used internationally for mobile telephone systems and is based on circuit-switching technology (as opposed to the sole use of packet-switching in GPRS) [427]. Cellular network operators have approved the use of GSM networks for AMI communications because they provide sufficient bandwidth, data rates, anonymity, and protection of data [412, 424]. At this point, 3G technology is a mature network with a completed theory and experience [412]. It is secured using various encryption technologies, but its security can still be a concern. Its communication rate is not reliably real-time [412].

More recently, the 4G and LTE standards have been developed. 4G was defined by the International Telecommunication Union (ITU) using many of the 3G standards. In 2007, the Third Generation Partnership Project (3GPP) completed its task of creating the LTE standardization [448]. The project's objective was to meet increasing requirements on higher wireless access data rate and better quality of service [448]. Subsequently, 3GPP immediately started a standardization process called LTE-Advanced for 4G systems [424, 448]. Because of its high reliability and low latency, LTE is suitable for NAN smart grid applications such as automated metering systems and distribution system control [424]. Furthermore, LTE offers opportunities to scale deployment because it is widely supported and its hardware costs are expected to improve [424].

3.2.3.2 WiMAX Networks

In complement to the cellular data networks described above, the Worldwide Interoperability for Microwave Access (WiMAX) standard was developed by the IEEE 802.16 working group to meet 3G standards and then later revised to meet 4G requirements [448]. It has been developed for "first-mile/last-mile" broadband wireless access as well as backhaul services in high-traffic metropolitan areas [448]. WiMAX is a communication protocol that provides fixed and fully mobile data networking. It has versions that work with licensed and unlicensed FCC frequencies that work in the 10–66 GHz and 2–11 GHz ranges, respectively [427]. WiMAX has a theoretical data rate of 75 Mbps and is designed for larger areas with a range of up to 50 km with a direct line of sight [410, 427]. As a standard, WiMAX offers interoperable microwave access [424].

The WiMAX architecture is a proprietary network, which comes with the benefit of complete control to utilities [424]. It is well-suited for use in a NAN due to its bandwidth and range [412, 424]. It offers efficient coverage and high data rates [424]. It also has low latency and relatively low deployment and operating costs [424]. These characteristics favor smart meter networking and are sufficient to support the real-time data transfers required for real-time pricing programs [424]. Disadvantages of WiMAX include a high initial infrastructure cost for radio equipment, which requires optimizing the number of station installations and quality of service requirements [424].

3.2.3.3 eIoT Perspectives on Commercial Telecommunication Networks

As eIoT continues to develop technologically, it is clear that commercial telecommunications networks will have an increasingly important role. They provide sufficient bandwidth for wide-area data transfer that allow them to be used for distributed smart grid applications such as AMI and DERs [106, 423, 424]. These networks are suitable for NAN, where they can connect peripheral devices to private area networks [424]. The LTE and WiMax standards also have the bandwidth and quality of service capabilities to support NAN-to-NAN (N2N) communications [106, 423, 424]. Beyond simply speed and quality of service, telecommunication networks and their associated operators offer grid operators and utilities an existing and cost-effective means for networked energy management. Furthermore, utilities (especially smaller ones with limited technical staff) have the opportunity to outsource maintenance and security upgrades in networks that are continually evolving with new generations of technology. This allows utilities to focus more on "core" business services [424].

Despite these many advantages, the integration of telecommunication networks into grid operations faces potential challenges. Cellular networks serve a larger customer market, which may result in network congestion or decreased performance [106]. Critical communications applications may not find cellular networks dependable in an emergency such as a storm or abnormal traffic situations [106]. Furthermore, although the speed of cellular networks continues to evolve, the number of mobile devices and their demands for data is also continually growing [425]. Grid operators, utilities, and telecommunication networks will have to work collaboratively to ensure that telecommunication networks have sufficient capacity to handle a continually evolving eIoT and its associated energy-management applications. In some cases, a utility may prefer its own private network to ensure quality of service and reduce monthly operating costs [106, 424]. It is also possible to develop hybrid utility-telecommunication networks so that congestion events do not interfere with emergency utility operation. LTE, for example, has the ability to operate either as a default or as a backup network [424]. Finally, from the perspective of power grid cybersecurity, a public telecommunication network is often perceived as a vulnerable point of operation [423]. Further work is required to bolster security on public cellular networks given their new role in eIoT energy management [423].

Finally, as telecommunication system operators face the strains of increased mobile and wireless device usage, an advanced, next-generation technology (5G) is needed [425]. Mobile-cellular subscriptions increased from approximately 109 million to 355 million between 2000 and 2014 [449]. As more devices become wireless, the telecommunications industry must address the physical scarcity of the radio frequency spectra for cellular communications, increased energy consumption, and average spectral efficiency while maintaining high data rates, seamless coverage, and a diversity of quality of service (QoS) requirements [425]. Heterogeneous networks may cause fragmented user experience, and so compatibility of these devices and interfaces with networks must be ensured [425]. 4G network data rates may not be sufficient for cellular service providers [425]. Instead, they must adopt new technologies as a solution for the billions, perhaps trillions, of active wireless devices [425]. 5G is expected to be standardized around 2020 [425].

3.2.4 *Local Area Networks*

In addition to grid operator, utility, and telecommunication networks, there is a growing need for LANs at the consumer's premises. Such networks use local area, often low energy, communication technologies to connect to a wide variety of devices in the home, commercial building, or industrial site [427]. These LANs also route information from peripheral devices such as smart thermostats and water heaters to energy-management systems and smart meters and monitors [410]. Local area networks are also often connected via smart meters and internet gateways to other "smart grid" actors such as electric utilities or third-party energy service companies (ESCOs). Such gateways enable customer participation in the utility's NAN applications such as prepaid services, user information messaging, real-time pricing and control, load management, and demand response [410].

Because LANs support a tremendous diversity of peripheral devices, they are also characterized by a diversity of standards and protocols. This section highlights some of the more emergent technologies including [106, 427]:

1. Wired Ethernet in Sect. 3.2.4.1
2. WiFi in Sect. 3.2.4.2
3. Z-Wave in Sect. 3.2.4.3
4. ZigBee in Sect. 3.2.4.4
5. Bluetooth in Sect. 3.2.4.5

A brief discussion of industrial networks is also provided (in Sect. 3.2.4.6) to address the specific needs of industrial sites.

3.2.4.1 Wired Ethernet

Ethernet is a dominant wired technology and it is widely used in residences and commercial buildings [450]. Almost all personal and commercial computers are equipped with an Ethernet port, and Ethernet connections are increasing among consumer entertainment equipment [426, 450]. Ethernet using an unshielded twisted pair (UTP) cable has four different supported data rates (10 Mbps, 100 Mbps, 1 Gbps, and 10 Gbps) that are covered by the IEEE 802.3 standard [450]. Although Ethernet has a high data rate, not all devices in private networks may be suitable for Ethernet connection. These devices may not have Ethernet ports, such as many home appliances, or are in environments that cannot support the power requirements or justify the cost of Ethernet [426].

3.2.4.2 WiFi Networks

WiFi networks are the natural wireless alternative to wired Ethernet. WiFi provides high-speed connection over a short distance [427]. The IEEE 802.11 standard defines various WiFi ranges and data rates [427]. Its optimal data rates span from 11 to 320 Mbps, and its optimal range spans from about 30 to 100 m [427]. WiFi is not meant for moving devices, and although not intended for metropolitan areas it has been extended to larger areas [427]. This is due to its support of personal devices on wireless internet access. WiFi is an IP-based technology and is widely used for a variety of electronic devices such as computers and mobile phones [426].

3.2.4.3 Z-Wave Networks

Z-Wave is an example of a proprietary wireless communication technology in LANs [426]. It is most suited for residences and commercial environments with low-bandwidth data transfers [426]. It is able to include device metadata in its communications and is easily embedded in consumer electronic products due to its low cost and low power consumption [426]. Unlike WiFi, it operates in the 900 MHz range and can be customized for simple commands such as ON-OFF-DIM for light switches, and Cool-Warm-Temp for HVAC units [426]. Z-Wave compatible devices can also be monitored and controlled by gateway access to broadband Internet [426].

3.2.4.4 ZigBee Networks

Zigbee can be used as an alternative to WiFi and Z-Wave [423]. It is often used in industrial settings [427]. ZigBee can cover about 100 m with a data rate of 20–250 kbps according to the IEEE 802.15.4 standards [412]. In applications that do not require large bandwidth, ZigBee offers a low-cost solution [412, 427]. ZigBee has real-time monitoring, self-organization, self-configuration, and self-healing capabilities [423]. It is also appropriate to eIoT applications because LANs can use it to create a mesh network of devices whose range and reliability increases as more devices are added [412, 426]. ZigBee devices are battery-powered and this may factor into the choice of network topology (star, tree, or mesh) [412]. In general, ZigBee has low power consumption and reliable data transmission [412]. However, since ZigBee devices are smaller, they tend to have limited internal memory, limited processing capability, and low data rates [412, 423].

3.2.4.5 Bluetooth Networks

The Bluetooth protocol was developed to provide point-to-point wireless communication such as between mobile phones and laptop computers [451, 452]. Currently, it shares the IEEE 802.15 standard with ZigBee technologies. Bluetooth operates in the unlicensed 2.4 GHz spectrum [427]. In addition to point-to-point capabilities, it can create meshed networks with a range of 1–100 m at data rates of up to

3 Mbps [412, 427]. Its range and low power consumption makes it suitable for local monitoring of devices; however, Bluetooth is vulnerable to network interference and offers weak security [412].

3.2.4.6 Industrial Networks

In addition to the above communication technologies, there exist a number of communication technologies that are specific to industrial applications. As has been mentioned several times in the preceding sections, LANs must offer multi-level security, be cost effective, comply with standards, provide reliable transmission, offer ease of access and use. Industrial networks have several additional requirements including predictable throughput and scheduling, extremely low down times, reliable operation in hostile environments, scalability, and straightforward operation and maintenance by plant personnel (who are not specialized in communication systems). Ultimately, these (often competing) requirements have led to a diversity of industrial networks. Some of the leading industrial networks include [453, 454]:

1. DataHighway Plus
2. Modbus
3. Highway Addressable Remote Transducer (HART)
4. DeviceNet
5. ControlNet
6. Ethernet/IP
7. LonWorks
8. AS-1, P-Net
9. Profibus/Profinet
10. Foundation Fieldbus
11. Ethernet

A detailed review of these technologies is beyond the scope of this work, however, the reader is referred to the following references [453–456] for an introduction to the topic. In the context of this work, these industrial networks form the communication layer of the "industrial Internet of Things" (IIoT) [457–459]. Naturally, as energy management becomes an increasingly important part of industrial operations, IIoT and eIoT will be viewed as overlapping and complementary development rather than mutually exclusive.

3.2.4.7 Perspectives on Local Area Networks

The wired and wireless networks described above perform the communication function in homes, commercial buildings, and industrial facilities. As eIoT continues to develop Ethernet, WiFi, Z-Wave, ZigBee, and Bluetooth networks are likely to continue to exist alongside each other [106, 426, 427]. In most cases, the most important role of these networks is to connect peripheral "smart" devices back to centralized applications, such as home energy monitors, home hubs, or utility-facing smart meters. Smart meters, in particular, can act as an interface between the LAN and the NAN [106, 414]. Such interface can serve several purposes including remote load control and the monitoring, and control of DER and EVs [414].

Beyond traditional fixed applications, local area networks must increasingly support mobile devices. Unlike a fixed network topology, a mobile device must identify the network in which it operates, as well as the identity and location of its peer devices in order to operate properly [460]. The integration of mobile devices into LANs necessitates networks with changing topology and algorithms that enable the real-time discovery and update of new devices [460]. Such applications raise questions of network security. Data exchange and interface interactions must be supported by trusted and secure devices that gracefully recover from failure [428]. The security risk of an untrusted device entering the network (or a trusted device being hacked) increases as the attack surface of the network increases. LANs are dispersed, highly fragmented, last-mile communication networks of the electric grid [426]. This heterogeneity of devices and communication channels make it difficult to protect from security breaches and data poaching.

In addition to network security, the fragmentation in LANs also complicates their interoperability [426]. Each of the communication technologies described above has its associated advantages and no one standard is likely to emerge for all applications [106, 426, 427]. One solution is to use the IP as a unifying translation layer across many different heterogeneous networks [426]. In such a case, each "smart" device must have a usable IP (v6) address. Beyond LANs, IP can also serve to improve the interoperability with other networks such as SCADA. IP and "middleware" can deliver data to utilities in readable formats [412]. For these reasons, IP is viewed as an integral part of the widespread development of eIoT.

Finally, it is clear that communication networks will continue to require many thoughtfully developed technical standards. As communication networks are advanced, it is important to create protocols that:

1. Transmit data within a relatively small (private) area
2. Transmit data back to a central location
3. Provide backward compatibility to 2G, 3G, 4G, and LTE standards

Successful implementation of these open standards requires engagement of hardware and software companies in both the electric power and telecommunications sectors [132].

3.2.5 *IoT Messaging Protocols*

The previous sections have covered eIoT communication technologies that enable devices to form machine-to-machine networks using various radio technologies. For LAN, these may include Zigbee, Z-Wave, WiFi, or Bluetooth. This section now covers the messaging protocols that are used over communication networks. The messaging protocols discussed here include:

1. eXtensible Messaging and Presence Protocol (XMPP)
2. Advanced Message Queuing Protocol (AMQP)
3. Data Distribution Service (DDS)
4. Message Queue Telemetry Transport (MQTT)
5. Constrained Application Protocol (CoAP)

3.2.5.1 Data Distribution Service (DDS)

The DDS is a message-passing service that provides publish/subscribe capabilities [461, 462]. DDS has been used successfully to provide scalable and efficient applications within the LAN [461, 462]. This service is used for real-time M2M communication. Its architecture does not involve a broker thus making its communication a distributed service [461, 462]. DDS was developed to support any programming language and it is the only standard messaging application programming interface (API) for C and C++ [463]. Its publish/subscribe wired protocol allows for interoperability across various programming languages, platforms, and implementations [463]. It provides a quality of service (QoS) for different behaviors [463] but there have been suggestions to leverage the good features of DDS and MQTT to provide a more flexible QoS IoT applications [462].

3.2.5.2 Message Queue Telemetry Transport (MQTT)

IBM's MQTT is optimized for centralized data collection and analysis through a broker [462, 464]. It offers an asynchronous publish/subscribe protocol that is based on a transmission control protocol (TCP) stack [464]. Usually a client sends information to a broker or a subscriber elects to receive messages on certain topics [464, 465]. It provides three QoS options [461, 464]:

1. Fire and forget (no response necessary)
2. Delivered at least once (acknowledgement needed, message received once)
3. Delivered exactly once (ensure delivery exactly one time)

MQTT has been designed to have low overhead and is suitable to IoT messaging as no responses are needed most of the time [464]. The system may require username/password authentication especially for brokers and this is achieved through secure socket layers (SSL) /transport layer security (TLS) [464, 466].

3.2.5.3 Constrained Application Protocol (CoAP)

The CoAP was designed by the Internet Engineering Task Force (IETF), and is based on HTTP making it interoperable with the internet [467]. It offers a request/secure protocol that use both asynchronous and synchronous responses [464]. It provides four types of messages [464]:

1. Confirmable
2. Non-confirmable
3. Acknowledgement
4. Reset

It also allows for a stop-and-wait transmission mechanism for confirmable messages and a 16-bit "Message ID" is provided to avoid duplicates [464]. Due to its compatibility with HTTP, CoAP clients can access HTTP resources through a translation system [464, 468]. It does not offer any security features [464].

3.2.5.4 eXtensible Messaging and Presence Protocol (XMPP)

XMPP was initially designed for messaging and has been widely in use for over 10 years. However, due to its age XMPP is starting to become outdated for some of the newer messaging requirements [464]. For instance, Google recently stopped supporting it [469]. XMPP runs on TCP and provides both asynchronous publish/subscribe and synchronous request/respond messaging systems. Given that it was designed for near real-time communication, XMPP is suitable for small and low-latency applications [464, 470]. It offers the specification of XMPP extension protocols to expand its functionality [464]. It has TLS/SSL built in for security purposes but does not offer any QoS [464]. It also uses XML which may cause additional data overhead and increased power consumption [464].

3.2.5.5 Advanced Message Queuing Protocol (AMQP)

AMQP came out of the financial industry [464]. It mainly uses TCP but can use other transport services as well. It offers asynchronous publish/subscribe protocols and has a store-and-forward feature that ensures reliability when service is lost [464, 471]. It provides three QoS [464]:

1. At most once (message sent once whether it is delivered or not)
2. At least once (message delivered one time)
3. Exactly once (message delivered only once)

Security is provided through TLS/SSL. AMQP may have low data rates at low bandwidths [464, 472].

3.3 Distributed Control and Decision Making

Thus far, this chapter has closely followed the generic control structure in Fig. 3.1. Section 3.1 highlighted the tremendous heterogeneity of network-enabled physical devices that are integrated across the electric power grid to measure and control primary and secondary variables on the supply and demand sides. Their deployment naturally inspired the development of multiple mutually coexisting communication networks. Section 3.2 differentiated these networks based upon their operator, traditional grid operators, telecommunication companies, and finally LANs belonging to residential, commercial, and industrial customers.

These two large-scale trends are transformative. No longer is the grid composed of thousands of centralized and actively controlled generators supplying billions of passive device loads. Rather, the centralized generation is complemented by distributed renewable energy that is often variable in nature. Furthermore, many of the passive device loads have become active and network enabled [45, 46]. The last step in the activation of the grid periphery is control and decision-making algorithms that serve to coordinate these devices to achieve balancing, mitigate line congestion, and meet voltage control objectives. Given the spatial and functional distribution of these devices, scalable and distributed control techniques that efficiently represent all the interactions are required to control and coordinate them, whether the interactions are collaborative or competitive [473].

In order to meet the challenges presented by the grid's physical transformation, the structure and behavior of the power system's operation and control must similarly change. Figure 3.14 shows a generic hierarchical control structure for a typical power system area. Passive loads are aggregated by a distribution system utility and passed to an independent (transmission) system operator (ISO) [20]. The ISO runs a wholesale day-ahead electricity market in the form of a centralized

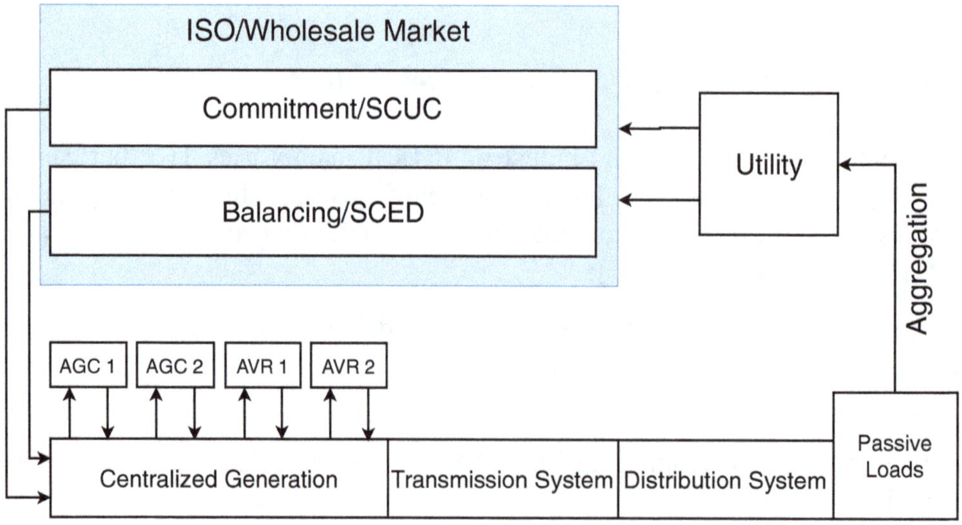

Fig. 3.14 A generic hierarchical control structure for a typical power system area

security-constrained unit commitment (SCUC) as well as a finer-grain "real-time" balancing market in the form of a security-constrained economic dispatch (SCED). These two market layers approximate the aggregated load at 1-h and 5-min intervals, respectively.

Decentralized automatic generation control (AGC) and automatic voltage regulation (AVR) use feedback control principle to adjust frequency and voltage at finer timescales (on the order of 1 Hz). Typically, each of these control layers is studied independently, often separating technical and economic analyses [15]. More recently, the Laboratory for Intelligent Integrated Networks of Engineering Systems (LIINES) has advanced the concept of "enterprise control" to simulate, design, and assess such a hierarchical control structure holistically [245, 246, 474–478]. An extended rationale for power system enterprise control has been published relative to the methodological limitations of existing renewable integration studies [15, 245, 246].

Such an approach must now evolve again to address the grid's physical transformation. The centralized optimization algorithms found in the market layers of the generic hierarchical control structure (in Fig. 3.14) do not scale and are unable to address the explosion of active demand-side resources at the grid periphery [15, 17]. Furthermore, the decentralized control algorithms found in AGC and AVR lack coordination beyond their local scope of control. For these regions, effective control algorithms that provide both scalability and wide-area coordination are necessary [479, 480].

Perhaps one of the key research areas in distributed power system control is in solving the optimal power flow (OPF) problem in a distributed manner [481–494]. Not only is this problem difficult to solve (by virtue of it being non-convex), it also consumes significant computational resources. Being able to solve the problem in a distributed manner allows for faster solutions to the OPF problem, and larger problem sizes. A common technique is usually based on augmented Lagrangian decomposition [493, 495, 496] such as dual decomposition [482, 497], the alternating direction method of multipliers (ADMM) [483, 484, 492, 494, 496, 498, 499], alternating direct inexact Newton (ALADIN) [485], analytical target cascading (ATC), and the auxiliary problem principle (APP) [486, 500]. The other common approach is based on decentralized solution of the Karush–Kuhn–Tucker (KKT) necessary conditions for optimality and gradient dynamics [487]. The ADMM is by far the most common of these techniques [488]. Other distributed control study areas include wide-area control problems, optimal voltage control, and optimal frequency control [501]. Despite extensive publications in this area, guaranteed convergence remains a concern for most of these approaches [501].

While the transmission system is likely to remain unchanged, the distribution system can implement two distribution system energy markets with distributed algorithms. Furthermore, eIoT devices have the potential to provide AGC and AVR ancillary services. In some cases, the communication networks described in Sect. 3.2 will be sufficiently fast to enable the distributed algorithms. In other cases, network latency will limit these implementations to decentralized control [502].

To that effect, the power systems literature has developed significant work on multi-agent system (MAS) distributed control algorithms. In MAS applications, agents are equipped with the ability to simplify decision making by allowing them to communicate with few of their immediate neighbors and make decisions that then inform higher-level decisions [503, 504]. This ensures that devices do not carry too much information, and allows for better coordination within the system [503]. Key MAS features such as modularity, scalability, reconfigurability, and robustness make them especially paramount to the realization of distributed control [505]. This section seeks to highlight some of the important outcomes of this research.

Perhaps the earliest works on multi-agent systems in power system research occurred at the turn of the century in the context of market deregulation. Then, it was recognized that as power system markets shifted from a single grid operator to multiple competing generation companies that such "genCo's" would deploy new "game-theoretic" bidding strategies to maximize their profit. Therefore, some of the first works on the applications of multi-agent systems to the power industry were focused on modeling electricity markets in a deregulated power industry [506–510].

At the time, most algorithms studied the effect of self-interested agents on auction market equilibrium with a particular focus on the unit commitment problem [511–514]. As such, these MAS frameworks were composed of a few mobile agents, generator agents, and a market facilitator who would oversee the market bidding process [515]. Game-theoretic strategies were also employed to investigate potential coalitions or cooperative strategies among different competing parties [516, 517].

Around the same time, various MAS approaches considered optimal cost allocation techniques to manage cross-border exchanges, be it through tie-lines, or cross-jurisdictional transmission lines [518–520]. These trends reflect the earliest MAS trends that set the stage for later applications in electric microgrids, demand response, and smart grids.

MAS applications later diversified to other aspects of power systems control and operations such as balancing, scheduling, line control and protection, and frequency regulation [509, 521–525]. As more renewable energy resources have gained prominence in grid operation, MAS frameworks, too, have shifted focus to the provision of ancillary services. A significant number of studies have considered system restoration under vulnerable system conditions, and later these approaches have been applied to microgrids with some penetration of variable energy resources. Usually, these MAS applications study only a single layer of either economic or technical control [32]. In some cases, a MAS economic layer was combined with a single physical layer [32]. Later on, MAS applications came to incorporate demand response at the microgrid and residential levels [526–529].

Agent-based and game-theoretic approaches have also been applied for cooperative and competitive demand-side management and microgrid control [530–537]. Grid level MAS applications have focused on the provision of ancillary services, and in some cases the parallelization of grid-level communication and control networks such as SCADA [528, 529]. Game-theoretic approaches such as cooperative and non-cooperative games have shown great promise in the design of distributed control strategies for demand-side management [473, 538]. However, given the dynamic nature of the smart grid, these works showed that a stable equilibrium was not always possible in the presence of faults and slow learning speeds [473].

Multi-agent electric market simulators were also advanced to help in the study of competitive electricity markets. One such simulator is the multi-agent system competitive electricity markets simulator (MASCEM) which combined agent-based modeling and simulation to study the dynamics of competitive electricity markets [539–545]. Continued research is required to design distributed algorithms that use game-theoretic principles and ensure robustness, stability, optimality, and convergence.

Another important application of multi-agent systems in power systems has been the control and energy management of microgrids. There, it was recognized that microgrids are often implemented in remote and potentially harsh environments. Their associated centralized controllers and energy-management software present a single point of failure [503, 546, 547]. MAS in contrast are fundamentally more resilient in that they can continue to operate in the face of certain types of disruptions. Such a functionality is enabled by a modular decision-making architecture composed of semi-autonomous agents that allows agents to be added and removed without the need to halt the entire system.

A modular architecture is particularly vital as the penetration of variable energy resources (VER) grows because it allows for other energy resources to be easily reconfigured to support microgrid operation [548]. For example, the ability to island part of the microgrid to allow it to heal is of paramount importance in the control of microgrids with a high penetration of VERs [548–550]. As a result, many MAS frameworks have studied self-healing mechanisms of microgrids [548, 551–555] and some have even demonstrated resiliency of such microgrids under several reconfigurations [551].

Recognizing the distributed manner in which microgrids are controlled, distributed MAS-based algorithms have also been proposed for various, usually, hierarchical microgrid control applications. These control applications include economic dispatch [556], load restoration [557], decision making [558, 559], and scheduling [560] to name just a few. There has also been significant research on the control strategies for microgrids in islanded operation [549, 561, 562] to ensure reliability within the islanded system. Naturally, a lot of attention has gone into designing and standardizing the informatic interfaces of multi-agent frameworks. These frameworks have been designed to closely follow IEC 61850, IEC 61499 [563], and IEC 60870-5-104 [564] as standard architectures for interoperability.

In the meantime, further research needs to ensure that agent groups can perform functions at or near real-time. Furthermore, more work is required to assess the performance of distributed algorithms with respect to optimality and its global behavior relative to centralized algorithms [479].

Despite this extensive MAS research in power systems, an important limitation has emerged. Much like what has happened with traditional hierarchical control structures in the transmission systems, these MAS research works generally only address one control layer at a time. Furthermore, there is a significant dichotomy between MAS that controls physical variables to secure grid reliability and those

Table 3.5 Adherence of existing MAS implementations to design principles [32]

	[1]	[2]	[3]	[4]	[5]	[6]	[7]	[8]
1	Model limited to lines & substations. No model for power generation & consumption.	Model limited to power generation, consumption & storage. No agents assigned to grid topology.	Model limited to power generation, consumption & storage. No agents assigned to grid topology.	Model limited to power generation, consumption & storage. No agents assigned to grid topology.	Model limited to power generation, consumption & storage. No agents assigned to grid topology.	Model limited to power generation, consumption, and lines. No agents assigned to buses, storage, RE, or dispatchable load.	Model limited to load and bus agents.	Model addresses all power system structural degrees of freedom.
2	One physical resource has many function blocks. Each function block is meant to be part of a larger control agent.	Each agent has a physical resource. Not all physical resources have an agent.	Some physical agents are included. Some centralized agents are included. No agents assigned to grid topology.	Some physical agents are included. Some centralized agents are included. No agents assigned to grid topology.	Each agent has a physical resources. No agents are assigned to grid topology.	Some physical agents are included. Some centralized agents are included.	Some physical agents are included. Some centralized agents are included.	1-to-1 relationship of physcial agents to resources.
3	Model limited to lines & substations. No model for power generation & consumption.	Model limited to power generation, consumption & sotrage. No agents assigned to grid topology.	Model limited to power generation, consumption & sotrage. No agents assigned to grid topology.	Model limited to power generation, consumption & storage. No agents assigned to grid topology.	Model limited to power generation, consumption & storage. No agents assigned to grid topology.	Model limited to power generation, consumption, and lines. No agents assigned to buses, storage, RE, or dispatchable load.	Model limited to load and bus agents.	Model addresses all power system structural degrees of freedom.
4	Does not address the aggregation of generators, loads, or power grid areas.	A grid agent is included as a single entity rather than an aggregation of multiple entities.	A microgrid manager agent is included as a centralized decision-making entity.	Centralized agents are included for centralized decision-making.	Does not address the aggregation of generators, loads, or power grid areas.	Centralized agents are included for centralized decision-making.	Centralized agents are included for centralized decision-making.	Centralized agents are included for centralized decision-making.
5	Only Line & substation availability. Generation/Consumption not included.	All agents are assumed to be online.	All agents are assumed to be online. Microgrid can operate in grid-connected and disconnected modes.	All agents are assumed to be online.	All agents are assumed to be online.	All agents except for central agent & grid agent can be unavailable.	Only Line & substation availability. Generation/Consumption not included.	All agents can be switched on/off.
6	Function block interactions exists between lines & substations but not with generation & loads.	Without agents assigned to the topology agents, there can be no coordination between energy and topology elements or between topology elements.	Without agents assigned to the topology agents, there can be no coordination between energy and topology elements or between topology elements.	Without agents assigned to the topology agents, there can be no coordination between energy and topology elements or between topology elements.	Without agents assigned to the topology agents, there can be no coordination between energy and topology elements or between topology elements.	Without agents assigned to buses, storage, RE and dispatchable loads, coordination decisions are limited.	Without agents assigned to other physical resources, coordinated decisions are limited.	Agent architecture does not include interaction between branches, buseses & energy elements.
7	No extraneous agent interactions have been added.	Supercondensator initiates all negotiations with other agents in a sequential fashion.	Introduction of multiple centralized decision-making agents likely to add extra agent-to-agent communication	Introduction of multiple centralized decision-making agents likely to add extra agent-to-agent communication	No extraneous agent interactions have been added.	Introduction of multiple centralized decision-making agents likely to add extra agent-to-agent communication	Facilitator acts as a centralized agent.	Introduction of centralized decision-making agent likely to add extra agent-to-agent communication
8	1-many cyber-physical relation but each function block is meant to be an automation object as part of a larger control agent.	Agents are assigned to PV, storage, and external grid. No agents for loads, lines, and substations.	Some physical agents are included. Some centralized agents are included. No agents assigned to grid topology.	Some physical agents are included. Some centralized agents are included. No agents assigned to grid topology.	Each agent has a physical resources. No agents are assigned to grid topology.	Some physical agents are included. Some centralized agents are included.	Some physical agents are included. Some centralized agents are included.	1-to-1 relationship of physcial agents to resources.
9	Fulfilled.	Fulfilled.	The use of centralized decision-making causes local information to be centralized.	The use of centralized decision-making causes local information to be centralized.	Fulfilled.	The use of centralized decision-making causes local information to be centralized.	The use of centralized decision-making causes local information to be centralized.	The use of centralized decision-making causes local information to be centralized.
10	IEC61850/61499 are used with function blocks. Does not consider FIPA-compliant agents.	Matlab simevents is used for the development of MAS. Not FIPA compliant.	FIPA Compliant JADE Agents	FIPA Compliant JADE Agents	FIPA Compliant JADE Agents	Matlab is used for the development of MAS. Not FIPA compliant.	FIPA Compliant JADE Agents	FIPA Compliant JADE Agents
11	SimPower Systems Model but specifics are not mentioned.	Physical model of a DC grid implemented in simulink.	Real-Time Digital Simulator/Power World Simulator as physical model.	Small-signal stability model implemented in Matlab.	Real-time diesel generator included.	Physical model of system implemented in Matlab	Physical system model of implemented in Matlab/Simulink w/o specifices.	Transient stability physical model implemented in Matlab.
12	Function blocks are intended as real-time execution agent for fast switching decisions.	None present.	None present.	f-V and P-Q controls implemented as real-time execution agents.	Governor control implemented as real-time execution agent.	Voltage and PQ control implemented as real-time execution agents.	None present.	Automatic Generation Control & Automatic Voltage Regulators implemented real-time execution agents.
13	Slower timescales are not considered	Coordination agents address energy-management functionality.	Coordination agents address energy-management functionality.	Middle level coordination and high level energy management agents address balancing and voltage control operation	Coordination agents address energy-management functionality.	Central agent address black start service.	Central agent addresses restoration service.	Coordination agents address energy-management functionality.
14	Since only one time scale is considered, function-block layer is flat.	Since only one time scale is considered, agent architecture is flat.	Energy management is considered for the day-ahead and real-time markets. Power grid dynamics are not.	Three layer agent hierarchy devoted real-time frequency control, voltage coordination and energy management.	Two layer control hierarchy: energy management & real-time frequency control.	Two layer control hierarchy: black start coordination & real-time control.	Since only one time scale is considered, function-block layer is flat.	Two layer control hierarchy: energy management & real-time control.
Decision	Fault Location, Isolation & Supply Restoration	Energy management	Energy management	Energy management, voltage control, small-signal stability	Energy management & Frequency Control	Black start coordination & Real-Time Control	Restoration Service	Energy management & Frequency Control
Implementation	IEC61499 Function Block Implementation w/ SimPower Systems Simulation	Matlab Simevents/Simulink Implementation	JADE Agents with Real Time-Digital Simulator/Power World Simulator	JADE Agents with Small-Signal Stability Matlab Simulator	JADE Agents with Real-Time JAVA simulation	Matlab Implementation	JADE Agents with Simulink/Matlab Simulator	JADE Agents with Transient Stability Matlab Simulator

[1]-(Zhabelova and Vyatkin, 2012; Higgins et al., 2011); [2]-(Lagorse et al., 2010); [3]-(Logenthiran et al.,2012; Logenthiran and Srinivasan, 2012); [4]-(Dou and Liu, 2013); [5]-(Colson and Nehrir, 2013); [6]-(Cai et al., 2011); [7]-(Khamphanchai et al., 2011);[8]-(Rivera et al., 2014a,b)

[1] Zhabelova and Vyatkin [566] and Higgins et al. [567]; [2] Lagorse et al. [568]; [3] Logenthiran et al. [569]; Logenthiran and Srinivasan [570]; [4] Dou and Liu [571]; [5] Colson and Nehrir [572]; [6] Cai et al. [573]; [7] Khamphanchai et al. [574]; [8] Rivera et al. [551, 552]

that control economic variables to implement distributed versions of traditional market structures. In a recent review, only eight works addressed multiple layers of technical and economic control [32, 565]. The same work assessed these works against 14 design principles that enable resilient eIoT integration. The result of the assessment is shown in Table 3.5. As a technology development roadmap, it identifies the need for further MAS development that:

1. Implements distributed control algorithms
2. Addresses both technical and economic control objectives
3. Addresses the multiple timescales found in the integration of variable energy, energy storage, and demand-side resources

Finally, it is important to emphasize that the effective implementation of distributed control algorithms requires access to real-time data, data filtering, coordination, and control [575]. Standards and architectures must be put in place as platform upon which such algorithms can operate. First, individual nodes must be equipped with the necessary memory and computing power for low-level control functions. Second, functional and control standards for devices must be agreed upon to ensure interoperability between platforms. Third, modularity must be applied as an integral design principle that facilitates the integration of ever-more sensors and actuators. Fourth, the computing capacity accorded to each node must match its functional requirements. Lastly, in a truly distributed system, each node must have all the information needed to re-initialize new nodes and initiate backup procedures in the case of failure [575]. These provisions facilitate the design and deployment of distributed control strategies.

3.4 Architectures and Standards

Fundamentally speaking, many of the discussions presented in this work thus far can be seen as large-scale architectural changes of the electric power system towards *decentralization*. In the original discussion on energy-management change drivers presented in Chap. 1, the deregulation of electric power markets was introduced. Figure 1.5 showed the deregulation or unbundling of electric power as a shift from centralized monopolies to multiple, decentralized, and competitive suppliers. Similarly, the integration of renewable energy and active demand response shown in Fig. 1.7 may be viewed as a fundamental change in the architecture of the physical electric power system itself. The role of centralized generation facilities is being eroded by distributed renewable generation. The previous section's discussion on distributed control algorithms addresses the shift from a more centralized control structure in Fig. 3.14 to a more distributed one. Together, these three separate discussions show that eIoT is entirely consonant with a decentralized architecture in regulation, operations timescale decision making, and the physical power grid.

These three large-scale architectural changes fundamentally change how power and information are exchanged throughout the electric power system. As has been discussed several times throughout this work, eIoT brings about the need for two-way flows of power and information where one-way flows were once common. The most common examples of these are at the grid periphery where distributed generation can cause power to flow back up the radial distribution system and where network-enabled demand-side resources both send and receive information as part of demand-response schemes. Such two-way flows change the way both cyber and physical entities in the grid interact with each other. Physical energy resources must accommodate the two-way power flows. In the meantime, "cyber" entities such as controllers, enterprise information systems, and organizations as a whole will have two-way informatic interactions with each other. For example, utilities of the future [30] may become "distribution system operators" that enable retail electricity markets. Consequently, their historical role as a load serving entity in wholesale electricity markets is also likely to change. These changing roles of "cyber" entities on the grid further indicates fundamental changes in the electric grid's architecture.

It is difficult to determine at this time what a future eIoT-enabled electric power system architecture will look like. It is clear that the grid cannot continue to operate in a centralized hierarchical fashion as it has in the past. On the other hand, a full transition to eIoT-enabled heterarchy and decentralization is improbable as well. Much research work still remains in order to achieve the holistic performance properties that centralized algorithms have already demonstrated and consequently centralized architectures are likely to endure in those conditions. The meshed communication networks (such as Z-Wave and Zigbee mentioned in Sect. 3.2.4) suggest distributed control architectures. However, their limited range similarly implies centralized nodes that aggregate peripheral devices and present them to the rest of the electric power system. Overall, the underlying trends that support eIoT remain strong and so decentralized and distributed control algorithms will take hold where possible. On a spectrum between total centralized hierarchy and complete decentralized heterarchy, the electric power grid's overall future architecture falls somewhere in the middle.

In recognition of these electric grid's evolving architectures, there have been efforts on both sides of the Atlantic to develop open and extensible architectures. Under EU mandate M/490, the Smart Grid Architecture Model (SGAM) was developed [26]. As shown in Fig. 3.15, it is a structured approach to modeling and designing use cases for power and energy systems. The architecture is organized into a three-dimensional framework consisting of domains, zones, and layers. These allow energy practitioners to structure the use case design in a clear and concise way.

Meanwhile, on the other side of the Atlantic, the Energy Independence Security Act (EISA) of 2007 describes severable favorable qualities of a future smart grid architecture including flexibility, uniformity, and technology neutrality [576, 577]. To that effect, the GridWise Architecture Council (GWAC) created its interoperability framework created its interoperability framework shown in Fig. 3.16 [27, 28, 578]. (This framework has often been nicknamed the "GWAC Stack" for simplicity.) Much like the SGAM, the GWAC Stack recognizes the need for multiple layers of integration in order to ensure interoperability, but does not add the dimensions of domains and zones. At the bottom, three layers ensure the interoperability of technical connectivity. When these layers are abstracted, they can form two informational layers that provide business context and semantic understanding. These layers may be further abstracted to form three organizational layers that address policy, business objectives, and business procedures. Both the SGAM and the GWAC Stack serve as the basis for the future development of

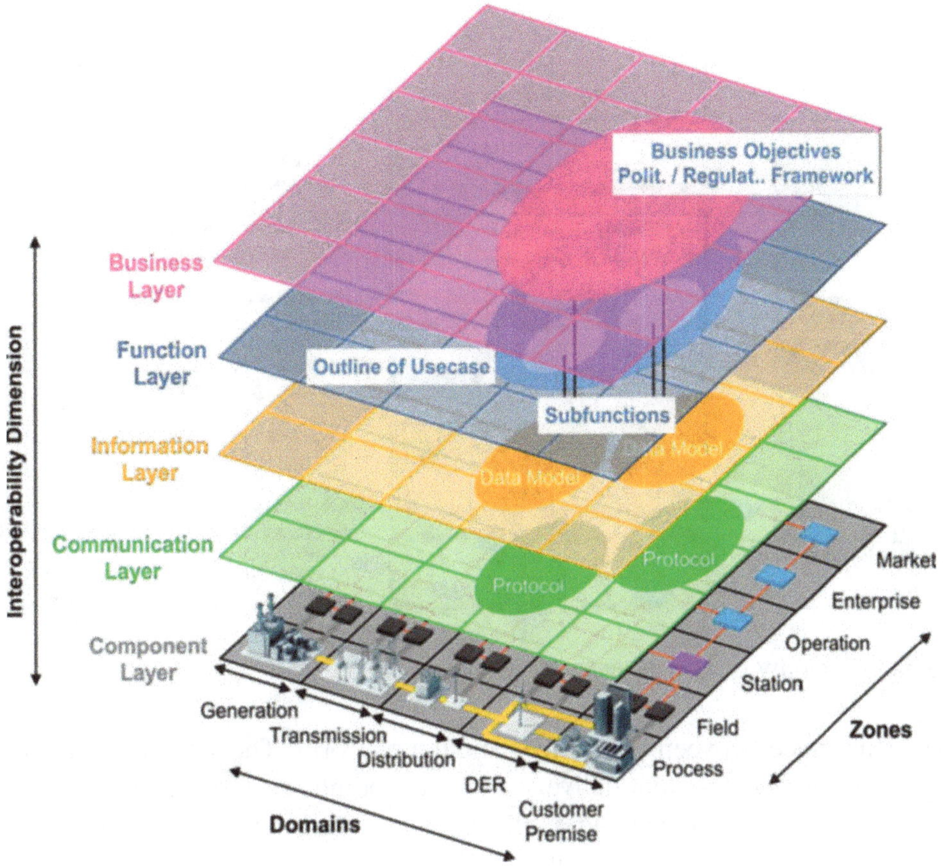

Fig. 3.15 EU mandate M/490 Smart Grid Architecture Model (SGAM) [26]

an electric power reference architecture that supports standard and interoperable implementations of eIoT.

In the meantime, there have been several efforts to develop commercial and quasi-commercial IoT platforms. Specifically, the OpenFog Consortium was launched in 2015 to spearhead the creation of an open architecture essential for creating IoT platforms and applications based on the fog computing ecosystem [579, 580]. The aim of the OpenFog Architecture is to accelerate the decision-making process of IoT sensors and actuators by bringing essential computation, networking, and storage closer to devices and reducing the latency brought about by all devices communicating directly with the cloud [579]. This architecture essentially serves as a middleman between the cloud and IoT devices and, thus, is not a replacement for cloud computing but rather complementary to the cloud [581]. The approach of bringing processing, that is, computation, storage, and networking closer to where the data is gathered is called fog computing, hence, the OpenFog Architecture [580, 581].

The OpenFog Architecture comprises of an OpenFog Fabric, OpenFog Services, devices and applications, and cloud services. The OpenFog Fabric is a computation platform on which services are delivered to all the devices [580]. The OpenFog

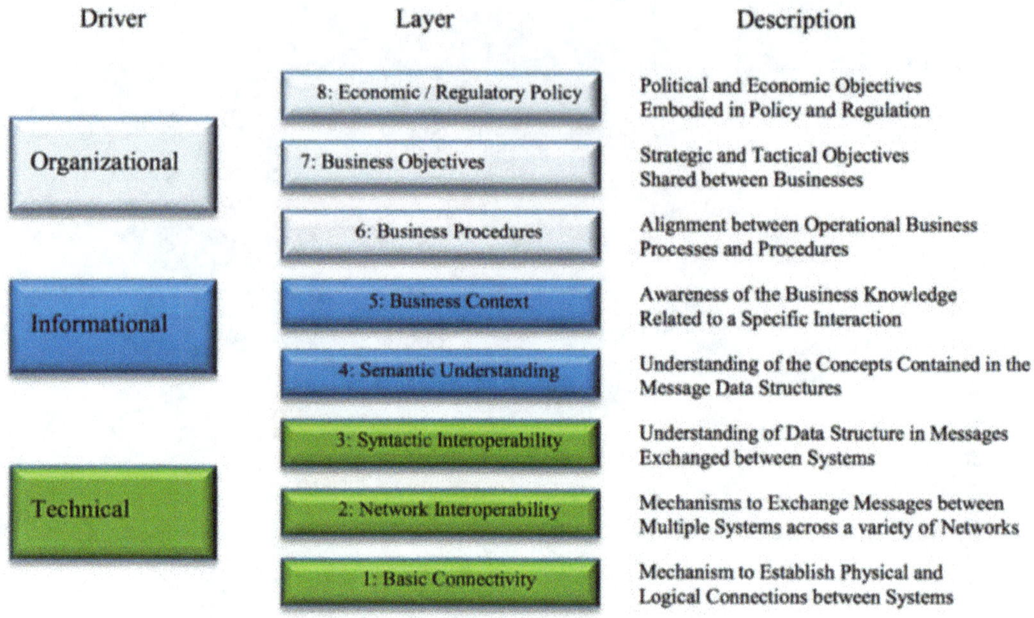

Fig. 3.16 The GridWise Architecture Council interoperability framework [27, 28]

Services interface between the devices and the platform. The services delivered by this platform include content delivery, video encoding, analytics platform to name just a few [580]. The device and application layer include sensors, actuators, and standalone applications running within or spanning multiple fog applications [579, 580]. Cloud services are available to be used for larger computational processes that later inform bigger decisions [579, 580]. The entire architecture is built to ensure the security of all communications and data. The OpenFog reference architecture is built upon eight pillars [579, 580]:

1. Security
2. Scalability
3. Openness
4. Autonomy
5. Reliability, Availability, and Serviceability (RAS),
6. Agility
7. Hierarchy
8. Programmability

Figure 3.17 illustrates the OpenFog reference architecture [580]. Recently, this reference architecture has been adopted as IEEE fog computing standard 1934 [580].

Other architectural standards are also provided by corporations such as Microsoft, Cisco, SAP, and Amazon. Amazon offers the Amazon Web Services (AWS) IoT Core which is a platform through which one can connect various IoT devices [582]. The AWS IoT comprises a device SDK that helps users connect and disconnect devices to the platform [582]. It provides broker-based publish/subscribe messaging through the MQTT, HTPP, or WebSockets Protocols

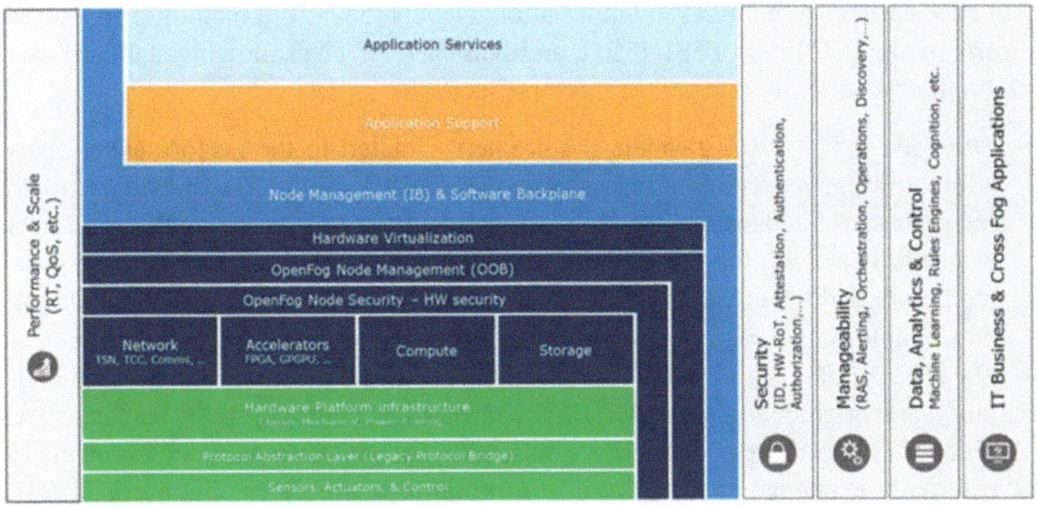

Fig. 3.17 The OpenFog reference architecture [27, 28]

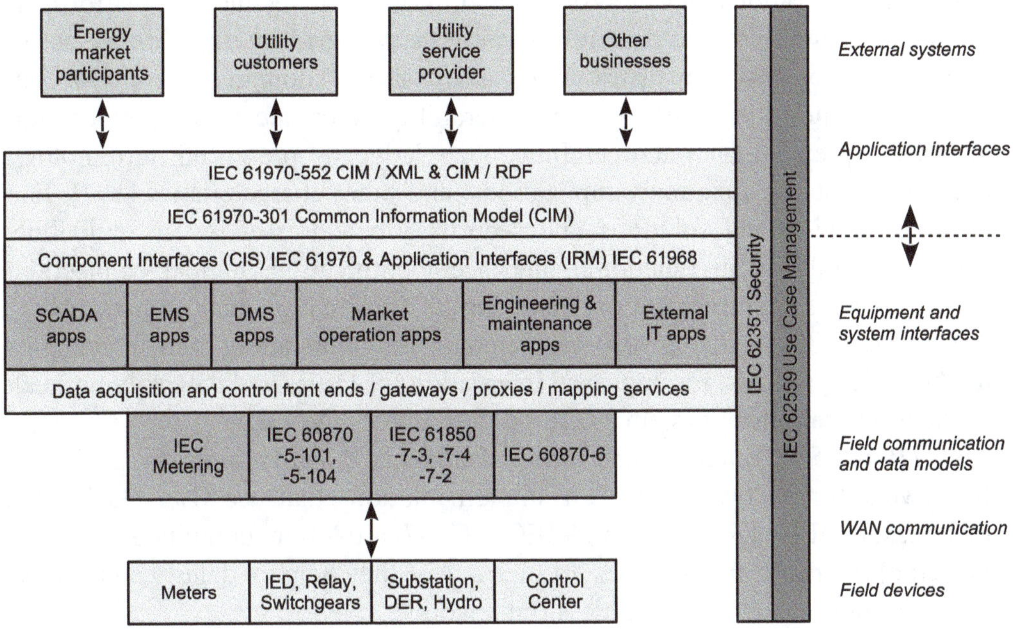

Fig. 3.18 An overview of important eIoT standards (adapted from [29])

[582]. The SDK supports C, Arduino, and JavaScript programming languages in addition to client libraries and a developer's guide [582]. SigV4 and X.509 certificate-based authentication is also supported by this platform [582]. Further discussion on this platform is beyond the scope of this book; however, more information on third-party IoT platforms can be found here for Amazon [582], SAP/INTEL [583, 584], Cisco [585], and Microsoft [586].

Consequently, the implementation of eIoT as automated and interoperable solutions rests upon a significant effort to develop effective standards. Beyond the communication standards mentioned in Sect. 3.2, several standards initiatives were launched early on at national and international levels [587–589] including concerted efforts by the IEC [590], IEEE [591], and NIST [577]. The following standards are highlighted as directly relevant [29, 592] (Fig. 3.18):

- *The IEEE 1547 Series* provide requirements related to the performance, operation, testing, safety, and maintenance of DERs [593]. The presence of an international standard was seen as a roadblock to the implementation of DG projects. That said, the standard does provide some technological flexibility for regulators at the local, state, and federal level [593]. The standard is intended to be technology-neutral and cover resources up to 10MVA.
- *The IEEE 2030 Series* establish interoperability as a basis for extensibility, scalability, and upgradeability [594]. IEEE 2030 defines interoperability as "the capability of two or more networks, systems, devices, applications, or components to externally exchange and readily use information securely & effectively" [593]. The standard is widely accepted as a pioneering document in architecture and interoperability in the electrical industry [414]. The standard uses perspectives from communications, power systems, and information technology platforms in the smart grid to provide design criteria for smart grid interoperability across generation, transmission, distribution, and customer domains [414, 594] The standard creates the smart grid interoperability reference (SGIR) model and supplies a premise for interoperability knowledge by presenting terminology, evaluation criteria, functions, applications, and other characteristics [594]. Furthermore, end-to-end solutions and security are addressed by its guidelines for interoperability in functional interface identification, logical connections and data flows, communications, and digital information management [594]. The IEEE 2030 standards help to maintain communications and information technologies progress, for improved integration for DERs and the evolving loads of the electrical power system [593].
- *IEC TR 62357 Seamless Integration Architecture (SIA)* aims to provide a framework for energy-related ICT implementations that use IEC TC 57. For this reason, IEC TR 62357 and IEC TC 57 are often combined to create (specific) reference architectures. In such a way, they help to identify and resolve inconsistencies and create seamless frameworks.
- *IEC 61970 Common Information Model (CIM)* specifies a domain ontology. In other words, it provides a kind of knowledge base with a special vocabulary for power systems. One goal is to support the integration of new applications in order to save time and costs. Another is to simply facilitate the exchange of messages in multi-vendor systems. The IEC offers an integration framework based on a common architecture and data model. In addition, the architecture is platform independent. The main application of IEC 61970 is the modeling of topologies.
- *IEC 61968 Distribution Management* extends IEC 61970 CIM for distribution management systems (DMS). These extensions relate in particular to the data model. The main use case is the exchange of XML-based messages in different DMSs.

- *IEC 62325 Market Communications* is also an extension to IEC 61970 CIM where the data model and messages are extended. However, the focus here is on market communication for EU and US-style electricity markets.
- *IEC 62351 Security for Smart Grid Applications* addresses ICT security for power system management with the goal of defining a secure communication infrastructure for energy-management systems with end-to-end security. This implies that secure communication protocols are specified in IEC 61970, IEC 61968, and IEC 61850.
- *IEC 61850 Substation Automation and Distributed Energy Resource (DER) Communication* focuses on communication and interoperability at the device level. The focal topics are:

  - The exchange of information for protection
  - The monitoring, control, and measurement
  - The provision of a digital interface for primary data
  - A configuration language for systems and devices

  This is implemented by:

  - A hierarchical data model
  - Abstractly defined services
  - Mappings of these services to current technologies
  - An XML-based configuration language for the functional description of devices and systems

- *IEC 62559 Use Case Management* deals with the steadily increasing system complexity associated with eIoT. In such a complex system, use cases help to structure and organize all relevant information for a technical solution. Therefore, in IEC 62559, five phases are identified for the development of use cases and the identification of requirements. Furthermore, a description template containing a narrative and visual representation of the use case is also provided.

Despite these many efforts in the development of eIoT architectures and standards, interoperability remains a formidable technical challenge to widespread eIoT implementation. In that regard, it is clear that the IEC, IEEE, and NIST will need to continue their efforts to enhance eIoT interoperability.

3.5 Socio-Technical Implications of eIoT

The previous sections have described the development of IoT within energy infrastructure in terms of network-enabled physical devices, communication networks, distributed decision-making algorithms, and architectures and standards. When taken together, it is clear that eIoT *fundamentally* transforms the relationship that "energy things" have with the information that describes them. The proliferation of sensing technology (described in Sect. 3.1) means that the *quantity* of information available to describe energy infrastructure will reach unprecedented levels. Beyond the quantity of information, the *type* of data will also diversify. Reconsider Fig. 3.2 on page 29.

Whereas much the electric power grid's data was associated with primary variables in the transmission system, Sect. 3.1.4 showed that this information will grow to include primary variables in the distribution system through smart meters. Furthermore, Sects. 3.1.3 and 3.1.5 showed that this information will grow to include secondary variables on both the supply and demand sides. These large and heterogeneous sources of data are also owned, generated, and transmitted by an unprecedented number of *stakeholders*. Reconsider Fig. 3.13 on page 55. The simultaneous presence of home area, neighborhood area, and wide-area networks implies that consumers will complement the role of utilities and grid operators as *generators of data*. As data is generated, natural questions will emerge as to the *ownership* of these data.

Finally, the extensive discussion on communication networks presented in Sect. 3.2 shows that the *transmission* of data will come to include telecommunication companies and private owners. Because eIoT fundamentally changes the role of information in energy infrastructure, there are two important socio-technical implications: privacy and cybersecurity. Both of these concerns are complex topics in and of themselves and cannot be extensively treated in the context of this work. Rather, this section seeks to provide an entry point from which more interested readers can more deeply investigate these topics.

3.5.1 eIoT Privacy

The proliferation of nearly ubiquitous eIoT data, particularly on the consumer side, raises important concern about consumer privacy. Reconsider Fig. 3.9 on page 45 which was mentioned in the context of home energy monitors that are able to infer the usage of individual home appliances based upon their electrical "signatures." While such information is very useful to a homeowner in the context of changing their own electricity consumption behavior, it can easily be used by other parties to infer a detailed picture of the homeowner's daily life including eating, sleeping, and leisure habits [595].

Beyond home energy monitors that point "inwards," smart meters are able to provide similar information (albeit at a lower sampling rate) directly to electric utilities. Naturally, many privacy concerns have erupted over this consistent flow of real-time data back to the utility because it can be mined with sophisticated data analytics algorithms to gain market power and potentially exploit the end-user. While the single example of smart meter real-time data flows is an important privacy concern, similar concerns can be found all over the eIoT landscape. The introduction of telecommunication and energy service companies as additional eIoT stakeholders further complicates privacy concerns and motivates the need for sensible policies that inform the rights and responsibilities of data generators, owners, transmitters, and users. The interested reader is referred to further works on eIoT Privacy [596–599].

3.5.2 eIoT Cybersecurity

The privacy concerns highlighted above gain further prominence in the context of cybersecurity. Returning back to Fig. 3.1 on page 3.1, every communication channel described in Sect. 3.2 has the potential to be compromised by an unintended or nefarious party. In some cases, such a party can *gather* data for potential gain outside of the grid. For example, a hacked smart meter could expose access to pricing information and communication networks in the home [276, 595]. In addition to the harm to end-users, the cost to the utility would be twofold. Not only could the utility be defrauded but it would also have to invest in fixing the problem [595].

In other cases, the unintended party can interject their data "upwards" to the control layer so that their associated algorithms have an incorrect picture of the physical world. For example, significant attention has been given to the impact of cyber-vulnerabilities of SCADA systems on the state estimators in operations control centers [600–602]. Similarly, nefarious parties can interject their data "downward" to the physical layer so that devices behave incorrectly. In both cases, the cybersecurity concerns become *cyber-physical* ones. For example, the automatic generation control feedback signal shown in Fig. 3.6 can be compromised so that the full control loop is no longer stable, consequently, placing the entire power generation facility at risk of failure [245].

These cybersecurity concerns become even more challenging in the context of the discussion in Sect. 3.2. Not only will eIoT communication networks be owned and operated by grid operators and utilities but they will also pertain to telecommunication companies and private end-users. While telecommunication networks have significant expertise in combating cybersecurity threats, private area networks are significantly more vulnerable. Consequently, significant attention will have to be given to the grid periphery to ensure that end-users are equipped with easy-to-implement cybersecurity solutions. The interested reader is referred to further works on eIoT cybersecurity [603–606].

Chapter 4
Transactive Energy Applications of eIoT

The previous chapters have situated the development of eIoT within an ongoing transformation of the electric power grid. In response to several energy-management change drivers, the grid periphery will be activated with an eIoT composed of network-enabled physical devices, heterogeneous communication networks, and distributed control and decision-making algorithms that are organized by well-designed architectures and standards. When these factors are implemented together properly, they form an eIoT control loop that effectively manages the technical and economic performance of the grid. This control loop is most consonant with an emerging concept of transactive energy (TE).

Definition 4.1 (Transactive Energy [607]) A system of economic and control mechanisms that allows the dynamic balance of supply and demand across the entire electrical infrastructure using value as a key operational parameter. ■

TE is commonly viewed as a collection of techniques to manage the exchange of energy in business transactions [47]. A utility, or any other private jurisdiction can implement TE between its various customers in industrial, commercial, and residential environments to manage DER technologies. TE applications incorporate the new eIoT-based activities for utilities, and industrial, commercial, and residential consumers. The result is better management of resources, successful integration of renewable energy, and increased efficiency in grid operations [47]. In many ways, TE is seen as an effective way to manage the technical and economic performance of various grid operations at all levels of control—commercial, industrial, or residential. As such, eIoT technologies directly support the implementation of TE applications.

This chapter discusses how aspects of the eIoT control loop from Chap. 3 are reflected in various TE applications across different layers of the electricity value chain:

- Section 4.1 discusses the role of TE in future grid applications and highlights some of the proposed TE frameworks.
- Section 4.2 presents a few motivational use cases for TE frameworks.

S. O. Muhanji et al., *eIoT*, https://doi.org/10.1007/978-3-030-10427-6_4

- Section 4.3 addresses the role of the utility and distribution system operators within the TE framework. This section also recognizes some of the challenges and opportunities presented by the implementation of TE.
- Section 4.4 examines several customer applications for TE and eIoT in commercial, industrial, and residential settings.

4.1 Transactive Energy

Transactive energy (TE) was a concept introduced by the GridWise Architecture Council (GWAC) to unite demand-side influences with wholesale markets, retail markets, and system operations [14]. GWAC is an organization that seeks to guide policy and facilitate the exchange of information in order to integrate information technology and e-commerce with distributed intelligent networks and devices [607]. A careful inspection of GWAC's definition of TE reveals that it is entirely consonant with the eIoT control loop. Not only does TE encourage dynamic demand-side energy activities based on economic incentives, it also ensures that the economic signals are in line with operational goals to ensure system reliability without resorting to override control [14]. It is this techno-economic nature of TE that makes it suitable to deal with the growing number of DERs and the current dynamic nature of consumer demand and energy market operations [14, 607].

TE is expected to offer increased efficiency to the power system and help maintain much needed reliability and security [607]. TE is, further, enhanced by its ability to engage both the technical and economic objectives of the grid in order to solve multi-objective control and optimization challenges [607].

As the grid evolves to accommodate more DERs, traditional grid control approaches must change to engage new grid stakeholders with more interactive control. DERs such as intelligent loads, storage, and distributed generation require more sophisticated control approaches than conventionally non-networked loads [607]. As more DER assets and their owners participate in the operational, economical, and semantic aspects of the grid [607], their activities must be optimally coordinated to align values and incentives among all stakeholders [607].

TE frameworks provide a systematic alignment of these incentives to favorably achieve grid objectives throughout central operations and peripheral additions. As a design rule, TE architectures must also account for the heterogeneity in the nature of transactions by providing the necessary definitions and guidelines. Recognizing the heterogeneous nature of operations provides the option to expand both the number and types of applications that can be added or removed from TE platforms. Consequently, heterogeneous operation includes making economic decisions that depend on local factors such as the levels of smart metering integration and DER penetration in the region [608]. Future TE development will rely on clear definitions of the transacting parties, the type of information to be exchanged, the transaction terms, what is being transacted, and the transaction mechanisms used by the system [607].

Recognizing that there is no "one size fits all" solution for interactions between the participants of the grid's techno-economic control loops [609], various groups have come forward to provide guidance in designing TE systems. The Transactive Systems Program (TSP) by the US Department of Energy aims to develop TE designs that offer "systematic, scalable, and equitable approaches for managing energy system operations [610]." The goal of TSP was to test existing TE designs to find an approach that is best-fit for the grid's multi-objective optimization problem. The program provides test cases and data sets for evaluating TE applications. It also outlines the criteria and procedures for measuring the performance of TE systems focusing on critical system behaviors such as scalability, optimality, and convergence [610]. Transactive mechanisms are key building blocks to energy exchanges, since each mechanism describes a value-based negotiation for energy flow between entities [610].

Recognizing the inter-timescale and multi-layer couplings of various grid operations, TSP analyzes mechanisms across varying timescales and layers of the energy system [610]. In addition, this program emphasizes the importance of creating the necessary interfaces to allow for communication, and interactions between various TE platforms as well as distributed control platforms [610]. It also stresses the need to clearly define any given TE platform to facilitate the transparent identification and comparison of TE frameworks [610]. TSP serves the key role of ensuring that TE platforms are assessed based on their value and overall contribution to the performance of the energy industry.

Another TE framework is the transactive energy market information exchange (TeMIX). TeMIX is a non-hierarchical methodology to support automation in energy transactions and decentralized control for the smart grid [47]. It is a subset of the Organization for Advancement of Structured Information Standards' (OASIS) for TE [47]. Essentially, TeMIX is a general marketplace for parties to interface in energy and energy transport transactions, with call and put options for both. Uniform information exchanges across DER component types occur in a TeMIX network for quotes, tenders, and transactions [47]. TeMIX allows for involved parties to carry out transactions without the intervention of any central authority thus removing any hierarchies. Transactions of energy and energy transport can occur between parties in retail and wholesale markets as well as between parties in different wholesale markets, a factor that is enabled by the standardized information exchange among all parties [47]. This simplifies interactions significantly by allowing exchanges across all parts of the electricity value chain. It is important to note that TeMIX is most useful in a smart grid context where customers are assumed to have smart meters, smart HVAC, and smart PEV charges [47].

Overall, TeMIX is a framework for automated interactions with the grid-periphery, consumer devices with distribution grids, transmission networks, and central generation and storage [47]. It simplifies the billing and settlement process for all consumer classes and DERs. Frameworks, like TSP and TeMIX, are important when planning transactions, since any modification to existing structures should undergo scrutiny from the perspective of holistic grid functions [609].

In addition to these TE frameworks, there have been several implementations and demonstrations of TE at the grid periphery in the past few years that have helped validate the TE framework for smart grid control. These demonstrations include the Olympic Peninsula Project, the American Electric Power (AEP) Ohio gridSMART® project, and the Pacific Northwest Smart Grid Demonstration (PNWSGD).

The Olympic Peninsula Project (OPP) was initiated in 2004 by the Pacific Northwest National Laboratory (PNNL) to test distributed dispatch, based on energy and demand price signals with automated, two-way communication between the grid and DERs [611]. This project implemented the GridWise concept which is a TE term coined by PNNL to describe a future-looking grid management system that uses smart devices and real-time communication [611]. GridWise technologies are a part of "non-wires solutions" (NWS) that are meant to provide alternative solutions to energy infrastructure issues due to growing load without having to build new transmission [612].

The Olympic Peninsula Project was carried out in Clallam County, the City of Port Angeles, and Portland, and served municipal, commercial, and residential loads. The project controlled a 150-kW water pump capacity between two stations, 175 and 600-kW generators, and 112 DR homes with two-way communication support in electric water and space heating [611]. Monetary incentives were used to control the DG suppliers and DR households. PNNL observed the DERs in this system through a dashboard that combined the resources as a common virtual feeder [611].

The main goal of the Olympic Peninsula Project was to assert the importance of intelligence at end use; that is to show that activating the grid periphery improves both the operational and economic efficiency of the grid [611]. This goal was guided by several sub-goals that include [611]:

- Show that DERs can provide multiple benefits through economic dispatch delivered by a shared communication network,
- Understand the individual and collective performance of DERs in near real time,
- Analyze the incentives and incentive structure for DER control and customer participation.

These goals not only helped study the value of active DER participation in energy markets but they were also a test of the effectiveness of current market practices [611]. Data from the system was collected for about a year (from early 2006 to March 2007) and were fundamental to the project's findings. The data provided unambiguous evidence that DERs could bid into the electricity market as a non-wire solution, and that these technologies could be applied at a larger scale [611]. Besides ascertaining the willingness of consumers to participate in DR given price incentives, this project provided a few key lessons for large-scale implementation of TE. In terms of increasing the number of participants, this project demonstrated that user-friendliness of the DR program or ease of participation were imperative for DR. As for grid operators, the ease of use relied on the availability of visualization dashboards that were developed and tested throughout the project.

The second project is the American Electric Power (AEP) Ohio gridSMART project. It focused on the deployment of advanced DR infrastructure in Columbus, Ohio [613]. The project embarked on infrastructural renewal by deploying advanced equipment such as smart meters, distribution automation circuit reconfiguration (DACR), voltage control and optimization from volt VAR optimization (VVR), and enhanced communication for consumer programs [613]. The project spanned 3000 miles of distribution lines, 16 substations, 100,000 residential consumers, and 10,000 commercial and industrial customers [613].

Given that no AMI meters had been installed in the region prior to the project, 110,000 m had to be installed to allow two-way communication between participants [613]. In addition to AMI, this project included cyber-security and interoperability requirements that involved comprehensive system improvement for both new and legacy systems [613]. The benefits of this program were numerous and provided a lot of insight for DR programs and grid operators. First, the AMI systems allowed for faster connections, remote-service usage, and improved billing accuracy. Second, automated circuit reconfigurations and smart metering infrastructure reduced the number of outages which in turn reduced field visits and manual meter readings. Furthermore, AMI could locate potential equipment failures to preempt outages and make the maintenance process more proactive.

The most notable benefits of this project were in consumer and pricing programs. In addition to smart meter installations, the project offered six programs that provided consumers with data on their energy usage and allowed consumers to respond to real-time price signals [613]. The real-time pricing with double auction (RTPda) was an experimental pricing program that was especially successful at allowing consumers to shift energy consumption according to fluctuating energy prices. Approximately, 250 consumers successfully participated in this program.

Another noteworthy benefit of this project was in the cyber-security and interoperability efforts. As a result of these efforts, multiple advancements were made to improve the security and interoperability of smart grid devices. The Cyber Security Operations Center (CSOC) was created to monitor and test the AMI system. Threat information was also shared with peer utilities and governments [613]. The CSOC was able to secure and validate the two-way communications from utility-owned networks through to the consumer home-area networks using penetration and interface testing [613]. Additionally, consumer data was protected with extensive and dedicated resources at a high level of security [613]. The CSOC continues to pursue efforts to ensure system security as well as interoperability in future deployments.

Like most projects, this demonstration was not without its challenges, and modifications will be required for any future deployments. The key challenge was in the deployment of new equipment. It was often costly, involved multiple maintenance team trips, and suffered equipment and communication system failures [613]. Despite these challenges, the program was an overall success; especially in creating awareness through community outreach programs and education [613]. The state of Ohio hopes draw from the lessons learned in Phase 1 and move to Phase 2 deployment where communication modules will be added to smart meters to enable DR and enhanced market participation [613].

The Pacific Northwest Smart Grid Demonstration (PNWSGD) by Battelle was arguably the world's first transactive coordination system [614]. This project was deployed in December 2009 and ended in 2015, funded by the DOE [614]. This project, in particular, exceeded the other two in both extent and complexity. It spanned multiple states and utilities, and included at least 55 smart grid systems [614]. Additionally, 25 out of 55 of the participating smart grid systems contained DERs of both supply and demand [614]. The cost and amount of electricity was negotiated to meet local and regional objectives, address renewable generation intermittency, and shape consumer loads [614]. Regional response was coordinated across 11 utilities, and a highlight of the project was the wide-scale connectivity between transmission, distribution, and home-area network systems. The demonstration successfully collaborated with dynamic endpoint responses to achieve conservation, reliability, responsiveness, and efficiency goals [614].

The tests in the PNWSGD were organized into three categories meant to bolster grid functionality [614]. First, several installations were made to contribute to improved energy conservation and efficiency [614]. Second, transactive assets were installed to respond to signals from the project's transactive system [614]. Third, these systems were tested for improved reliability in the distribution system [614]. These objectives of conservation, transactive response, and reliability were often investigated simultaneously at test sites [614]. A primary objective of the PNWSGD was to create a foundation for future smart grid advancements [614]. This objective was to be accomplished by creating an interoperable infrastructure to manage DR programs, DERs, and distribution automation in a system that could be validated through analysis [614]. This infrastructure combined generation, transmission, distribution, and load assets that were owned by utilities and customers across a five-state area [614].

An important focus for the project was data collection and analysis of the demonstration's costs and benefits for customers, utilities, and regulators [614]. The findings from the data provided potentially influencing testimonies for standards and methodologies for TE systems [614]. The project worked towards a future smart grid that is secure, scalable, and interoperable in regulated and non-regulated environments across the nation [614]. The transactive system was successful in connecting diverse, dynamic endpoint assets to the transmission system [614]. The report also noted that future applications of this system may further distribute its automated control responsibilities among distributed smart grid actors and devices [614].

Despite these successes, significant problems occurred with the consistent and accurate reporting of data [614]. Battelle expressed concern for utilities' ability to handle the large quantities of data that are produced by a smart grid system [614]. Future TE applications require better tools for confirming data accuracy. Furthermore, these applications must proactively identify and correct faulty sensing equipment that can introduce bad data into the system [614].

Together, these three TE demonstration projects have provided key insights into the opportunities and challenges of developing and deploying TE platforms. First, it is clear that TE systems must engage secure physical and cyber technologies to enable transactions. Second, these technologies must be interoperable so that devices with different functional characteristics can connect and communicate. Given that TE engages a diversity of systems, interoperable interfaces must allow transactive systems to operate across multiple timescales and enable event-driven operations [607]. Standardized interfaces must be constructed at the intersection of exchange mechanisms regardless of whether individual devices choose to play a transactive role [610]. Third, physical devices such as metering and telemetry devices must have the capabilities to accurately record and attribute energy flow measurements for the appropriate DR compensations [609]. In accounting devices, wholesale and retail services must be compatible to interoperate, yet also separable to prevent double counting for participants in multiple DR programs [609].

Since these TE demonstration projects, "blockchain" has emerged as a new internet encryption technology that enables distributed pricing [615]. Blockchain is a distributed cyber tool for communicating unique information publicly and securely [615]. Distributed, shared data repositories are protected from interference through encryption so that there is no need for extraneous bodies to enforce security [615]. At its core, a blockchain creates a "distributed ledger" as an immutable public record of transactions in a computer network [615] and entirely eliminates the need for a middleman. Transaction rates are determined by the size of distributed data sets, or "blocks," and the time interval for which the chain of data sets is periodically synchronized [616].

TE frameworks and enabling technologies are a force of decentralization that empowers DER management across energy customers. As a technology, blockchain shows great promise in enabling decentralized and distributed exchanges in TE applications. At the moment, blockchain protocols face scalability constraints that may slow transaction rates [616]. Nevertheless, blockchain has emerged as a technology that is integral to future TE applications.

In conclusion, TE platforms and applications are at the core of eIoT deployment and adoption. In the next subsection, the techno-economic control of TE is discussed in reference to its applications in industrial, commercial, and residential domains. The components of eIoT systems complement the high-level discussion of TE applications.

4.2 Potential eIoT Energy-Management Use Cases

The potential impact of TE can perhaps be best illustrated in two theoretical use cases. In one case, members of a community collaborate to lower costs by changing a utility's point of sensing. In the other case, larger loads or producers bypass utility involvement through direct participation in wholesale electricity markets. In both cases, energy consumers are able to make money by altering their relationship with utilities. These two eIoT TE use cases demonstrate how peripheral actors can engage in energy arbitrage with the help of present and future technologies. Opportunities for generators and consumers at the edge of the grid are presenting themselves in areas where price does not accurately represent the balance of supply and demand. Technological advancements in IoT enable peripheral actors to take action and exploit these imbalances in energy market prices. With eIoT, consumers and prosumers willing to form an aggregation can be set up to engage in energy arbitrage.

As first discussed in Sect. 1.3 and illustrated by the "duck curve" in Fig. 1.6 (on page 10), distributed power generation is expected, in the not too-distant future, to drive a surplus of energy compared to consumption during the same time [4]. Solar generation, in particular, is driving this trend, since its generation is limited by the hours of sunlight [4]. A glut in energy production during peak daylight hours does not necessarily coincide with consumers' energy demand [4]. The energy available on the grid during the surplus is sold at a low price, and sometimes at no cost. Hence, as prosumers inject their electricity into the grid, the value of this electricity falls, and so does the compensation received from utilities. If an oversupply occurs, utilities may curtail generation or bar the electricity from entering the grid. In most systems today, the retail price of electricity to consumers does not reflect the turbulent pattern of electricity supply [43, 44]. However, with implementation of TE systems, consumers can take advantage of lower energy prices.

Several assumptions are made to best present these use cases and to help guide the discussion:

1. It is assumed that eIoT technologies will be installed to the extent that sensing networks may adequately measure and process local consumption in real time.
2. A flow of pricing information from the electricity market to the periphery is available for consumers to react appropriately.
3. A connection to the market for energy flow and exchange is measured.
4. A platform to coordinate power data with pricing data is available to synthesize prosumer revenues and costs.

The eIoT technology trends described in Chap. 3 make these assumptions reasonable for the near future.

4.2.1 An eIoT Transactive Energy Aggregation Use Case

One interesting eIoT TE use case is based on the premise of changing consumers' relationship with a utility through aggregation. Consider Fig. 4.1. On the left, a conventional apartment building with rooftop solar consists of several apartments whose tenants act *individually* as conventional consumers to the local electric utility. Electricity consumption in each apartment is individually monitored with smart (residential) meters and the utility bills consumers accordingly. On the right, two important changes are made. First, the tenants of the apartment building

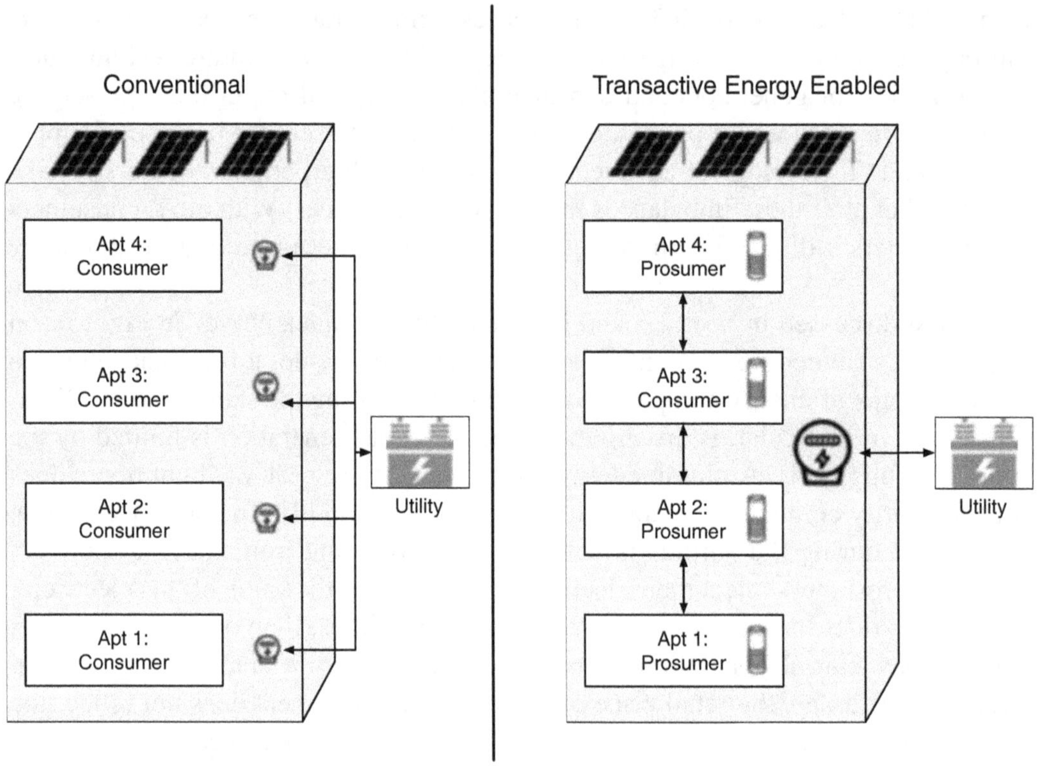

Fig. 4.1 A use case comparison between a conventional and an eIoT transactive energy-enabled apartment building

now act *collectively* as a single commercial *prosumer* to the local electric utility. Consequently, the many smart (residential) meters are replaced by a single smart commercial meter. Second, each prosumer purchases a TE-enabled smart home hub that allows each tenant to buy and sell electricity from other building tenants in real time.

The financial impacts on the utility and the tenants can be calculated. If the building as a whole consumes 2000 kWh at a rate of 0.1$/kWh and it generates from solar 1200 kWh which are sold back to the grid at $0.08/kWh, then the utility's total revenue for the conventional case is

$$\text{Utility Revenue} = (2000\text{ kWh}) * (0.1\$/\text{kWh}) - (1200\text{ kWh}) * (0.08\$/\text{kWh}) = \$104 \tag{4.1}$$

$$= \$200 - \$96 = \$104 \tag{4.2}$$

Collectively, the tenants spend $200 on electricity consumption and receive a $96 credit for their solar generation. In contrast, in the transactive energy case, the tenants with rooftop solar offer their solar generation at an average rate of $0.09 kWh to encourage their neighbors to shift their electricity consumption to daylight hours. Consequently, no solar generation is exported back to the grid. The utility's total revenue for the transactive energy case is

$$\text{Utility Revenue} = (800\,\text{kWh}) * (0.1\$/\text{kWh}) = \$80 \tag{4.3}$$

Consequently, the transactive energy case shows a $24 reduction in the utility's revenue! Even more interestingly, the tenants now spend only $188 as opposed to $200:

$$\text{Tenant Payment} = (1200\,\text{kWh}) * (0.09\$/\text{kWh}) \tag{4.4}$$

$$+ (800\,\text{kWh}) * (0.1\$/\text{kWh}) = \$188 \tag{4.5}$$

Finally, the tenants with rooftop solar now receive $108 as opposed to $96:

$$\text{Solar Tenant Credit} = (1200\,\text{kWh}) * (0.09\$/\text{kWh}) = \$108. \tag{4.6}$$

While this specific case may appear ideal, it is illustrative. In the TE case, the presence of solar generation provides an incentive for greater competition that ultimately benefits all the participating prosumers while simultaneously eroding the utility's billable energy. Because the tenants have collectively agreed to interact with the electric utility through a single commercial meter, the utility simply sees a decrease in the total amount of electricity purchased.

The eIoT TE aggregation use case above shows net social benefits due to several enabling factors:

1. The presence of prosumers with local solar generation that is, at times, inadequately compensated by utilities encourages the emergence of a transactive energy marketplace.
2. The solar generator's value proposition leaves local consumers at times over-billed by utilities.
3. The transactive energy marketplace is likely to be strengthened if there is a strong sense of community within the apartment building.
4. There exist nearly ubiquitous measurement, communication, and decision-making capabilities within the building to support the transactions. It provides price and quantity information for rational decision-making. The user-friendliness of these information technologies encourages greater adoption.
5. There exists a sparsity of measurement, communication, and decision-making capabilities between the building and the utility.

Naturally, if any of these factors is undermined, then the value proposition of the use case weakens. Of the five, only the last is directly within the utility's scope. Utilities and their associated regulators, for example, may choose to offer real-time retail electricity prices as a means of encouraging greater competition. In such a case, they would be encouraging TE at the distribution system level and not just at the building level. The alternative is that other TE buildings can emerge at the grid periphery. Furthermore, if such a trend were to take root, then large communities such as compounds and bounded neighborhoods might choose to do the same. In that case, a large enough TE microgrid could effectively form which bypasses a utility's services whenever it is convenient.

The application of the eIoT TE aggregation use case is already well suited for residential areas. Collaborations, such as the Brooklyn Microgrid project, embody aspects of this example and, in many ways, showcase the viability of peer-to-peer energy transactions [617, 618]. The Brooklyn Microgrid is a project that has brought consumers and prosumers to a virtual trading platform powered by blockchain to carry out energy transactions among themselves [619, 620]. This project, launched by LO3 Energy, provides a platform for consumers and prosumers to trade among themselves with the help of smart meters and blockchain technologies. A similar application is Power Ledger, a startup that was started in Australia, allows consumers to buy and sell renewable energy among themselves using blockchain [621]. In addition, Power Ledger intends to launch an asset-backed crypto token that will enable consumers or groups of consumers to share in the benefits of having renewable energy assets through trading in this token [621]. This approach would open the renewable energy market to a diversity of consumers and investors, hence, encouraging the growth of renewable energy systems [621]. Around the world, more and more people are starting to recognize the potential of peer-to-peer (P2P) energy transactions with some notable successes in Bangladesh, Germany, and New Zealand [619, 620, 622–624]. Beyond peer-to-peer applications, blockchain technology continues to support a growing number of applications in the energy industry. Recent studies have shown potential applications in cyber-security [625–627], multiple IoT applications [628–632], data privacy and security [633], and as a storage system for critical data [634]. Going forward, favorable regulatory measures might help advance peer-to-peer energy transactions such as those of the Brooklyn Microgrid. In customer applications such as this, TE implementation is primarily motivated by monetary incentives and the individual motivation to be more sustainable. Besides aggregation, energy usage can be modified at the source by adjusting times of use and consumption patterns.

4.2.2 An eIoT Economic Demand Response in Wholesale Electricity Markets Use Case

The second eIoT use case is based upon economic demand response (DR) as it is currently implemented in wholesale electricity markets. Consider Fig. 4.2. On the left is the same conventional apartment building. On the right is the same TE-enabled building which now acts as a single economic DR participant.

The building's conventional load profile is shown in Fig. 4.3a. For simplicity, assume that the building is relatively small compared to the peak load of the wholesale electricity market. Consequently, the building acts as a price taker because its bids have little effect on the locational marginal prices (LMPs) that clear the wholesale electricity market. Figure 4.3b shows the hourly LMPs for the full day. They are assumed to closely follow the trend of the "duck curve" mentioned earlier in Sect. 1.2.

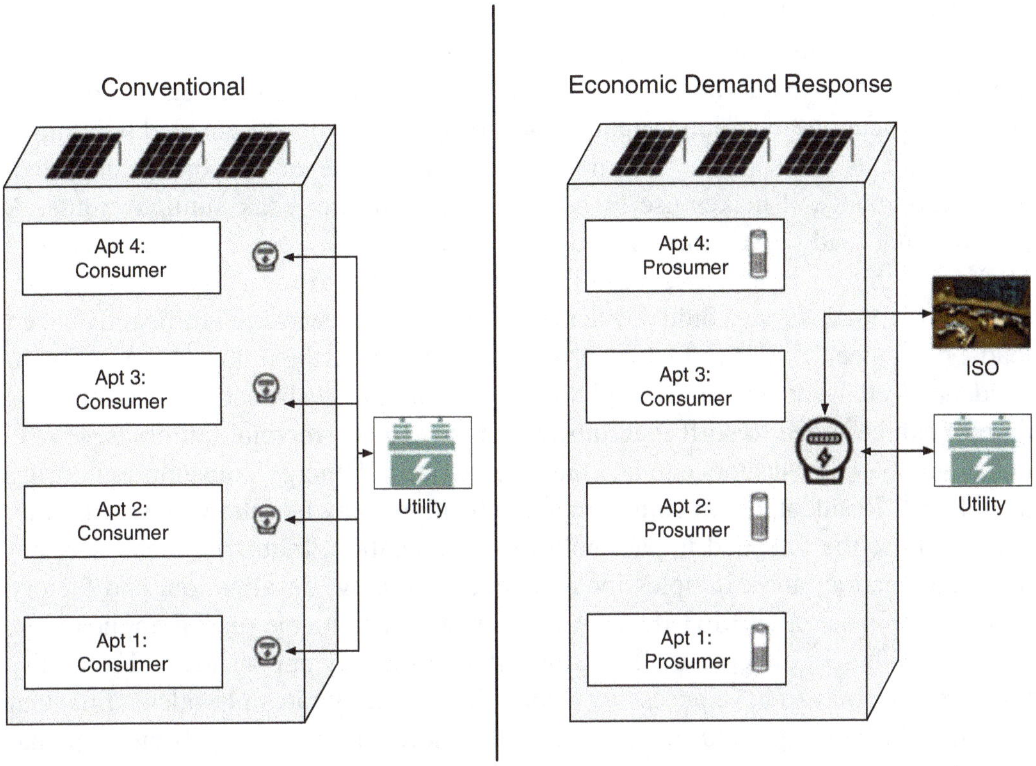

Fig. 4.2 A use case comparison between a conventional and an eIoT economic DR apartment building

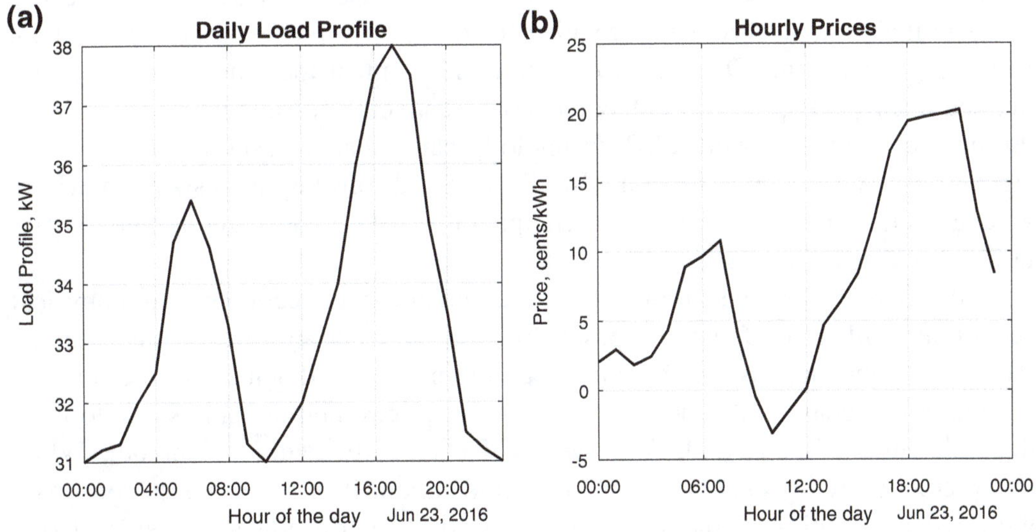

Fig. 4.3 eIoT Economic DR in wholesale electricity markets use case data: (**a**) On the left, the daily net load profile of the prior to demand–response incentives. (**b**) On the right, the hourly locational marginal prices (LMPs) experienced within the wholesale electricity market

The financial benefits for the transactive energy-enabled building can be calculated. As stated previously, the building's tenants pay \$200 when exposed to the retail rate. However, simply by entering the wholesale electricity market, they would pay $162 without shifting their behavior. This is because, on average, wholesale electricity rates are lower than retail rates. In such a case, the tenants have saved $38 but the utility naturally has lost all $200 because the TE-enabled building has effectively "cut out the middle-man." Now, imagine that the TE-enabled building is able to shift its loads so that it is no longer exposed to evening peak pricing and, more importantly, it makes use of negative LMPs during peak sunlight hours. A perfectly flat load curve would mean that the tenants now pay $134 for a savings of $66. In this case, as well, the utility has no access to the associated revenue. A flattened load curve could be achieved in multiple ways. Significantly sized loads, like a fleet of EVs or factory production, may have the required flexibility. In residences, eIoT-enabled home appliances (for example, dishwashers, washers, and dryers) can be timed to shift load during the day. In commercial buildings, HVAC units and hot water heaters can be controlled to curtail energy consumption during peak hours. Residential and commercial applications may be relatively small scale, but they have the intended impact with load aggregation. Industrial loads may not need aggregation, and examples include water pumping, desalination, and factory production. In all cases, eIoT devices and infrastructure enable the TE applications.

Again, this specific case is illustrative although it may appear ideal. The ability to aggregate so as to have access to wholesale electricity rates provides a financial benefit to the building's end consumers. Furthermore, the ability to participate in that market through economic DR allows the building to fill the troughs and shave the peaks of the duck curve. In both cases, this is financially beneficial [635]. Filling the troughs of the duck curve provides access to cheap and perhaps negative electricity prices. The peak shaving was not apparent in the case described above because the building's impact was small relative to the electric power system peak load. However, if economic DR were to become prevalent in the wholesale electricity market, then peak prices could come down and end consumers would benefit during these times as well. eIoT technology can enhance response to economic signals, and can ease coordination of production and consumption especially within an aggregate. The resulting direct participation in wholesale markets may bypass utilities; at least partially.

In the drive towards decarbonization, eventually carbon, economic, and physical accounting will align. If negative prices for renewable energy such as solar become the norm, then there is an economic opportunity to shift patterns of electricity consumption behavior. As the market adjusts to prices, and demand shifts to meet the imbalance of supply, duck curves will eventually begin to smooth. While this prediction relies on future eIoT implementation, it is nevertheless consonant with existing wholesale market practices. As the electric power system's market structures evolve to accommodate TE, it is clear that market facilitators will be required to coordinate new market procedures and entrants. Looking ahead, the question of who will take on this role remains an important component in the success of TE.

4.3 Applications for Utilities and Distribution System Operators

As seen in Chap. 3, the eIoT control loop is an electric power application that has the potential to transform the landscape of energy services for both consumers and grid operators. Furthermore, TE applications help create an empowered consumer base that is capable of making economically informed energy decisions that directly engage in energy markets. These factors put pressure on utilities to re-evaluate their approach to handling DERs and more likely reconsider the nature of their role in consumer applications. The two use cases discussed in Sect. 4.2 illustrate scenarios where utilities may face a future where consumers bypass their services partially or potentially altogether. This future scenario is not too hard to imagine especially with the DER innovations that are pressuring utilities to change their business-as-usual operations and increasing the accessibility of energy markets to consumers. The transition to transactive systems provides plenty of opportunities for utilities to take on energy-management services for customer DERs as well. However, there is no certain future for the overall transformation of the electrical power system especially regarding the role of utilities in consumer operations. Several questions are yet to be fully answered:

1. What will the transformation of utilities look like?
2. Will utilities take on the role of implementing TE?
3. What energy-management solutions for consumers will persist?

Concern for utility viability is not unique to today. The term "Death Spiral" once described the circuitous pattern utilities experienced in the 1980s of raising prices to cover costs, only to lose demand and make less profit [636–638]. Concerns about losing customers to distributed generation has revived the term, in that raising energy rates would lose profits for utilities by providing incentives for customers to generate their own electricity [637, 638]. While financial investors have found that this serious concern may be exaggerated, disruptive DER technologies and increased competition in energy markets have diminished utilities' abilities to seek rate increase in response to adverse economic environments [636, 638]. As a result, utilities may need to change their long-term strategy, as they did in the 1980s to deal with this potential "Death Spiral." The challenge of adjusting to disruptive eIoT technologies while simultaneously re-imagining their position in increasingly competitive markets makes the task for utilities much greater [637, 638].

The change drivers originally discussed in Sect. 2.1 are manifesting themselves into timely and pressing calls for action on the part of regulators and grid operators. For example, utilities in California are facing regulatory pressures to transform their businesses to accommodate DERs [4]. In the summer of 2016, the California Independent System Operator (CAISO) received federal approval for a Distributed Energy Resource Provider (DERP) tariff that allows aggregation between 500 kW and 10 MW of distributed energy to be submitted to the day-ahead and real-time energy markets as well as the ancillary services markets [639, 640]. This initiative not only poses technical challenges to CAISO but also calls for greater collaboration with utilities and any new market players willing to take on the role of managing DERs.

At present, CAISO has access to the transmission–distribution interface, while utilities own and control data between consumer-level metering and the distribution system [639]. As a result of this information gap, CAISO's DERP plan requires active collaboration with utilities. In addition, CAISO requires extensive network upgrades to address any operational concerns that may arise from this integration. If not planned carefully, it is possible that DER participation may not lead to reliable operation of the distribution system. Furthermore, without distribution data, CAISO may have to worry about larger effects aggregating up into the transmission system [639]. It is clear that the challenges described above span the technical and economic layers of grid operations. With the right investments, utilities could embrace new approaches that encourage the dynamic development of the grid and increase revenue in the process.

DERs create many new responsibilities for "distribution system operators" (DSOs) such as managing consumer data, and deploying new infrastructure such as advanced metering infrastructure, distributed storage systems, and EV-charging infrastructure [30]. With DERs, the role of utilities in operating the distribution grid becomes more complex because new suppliers and demand aggregators can emerge. Naturally, favorable regulations and tariffs are needed to promote the growth and adoption of DER technologies throughout the electric grid [30].

In addition to the production and investment credits for renewables, there have been new regulations favoring effective DER integration in market operations. In April 2016, the Federal Energy Regulatory Commission (FERC) put forward a Notice of Proposed Rule-making (NOPR) that required regional transmission organizations/independent system operators (RTO/ISOs) to revise their market rules to allow effective integration of electric energy storage into wholesale markets and the recognition of distributed energy aggregators as wholesale market participants [69].

The NOPR recognized that it was important to accommodate the operational characteristics of these DERs to allow them to participate competitively in wholesale markets [69]. This proposition was put in place in order to improve competition and encourage fairness in market rates by removing any potential barriers that hindered the effective integration of DERs [69]. As is currently the case, DERs may be hindered from participating in electricity markets due to the fact that the current market rules were specifically designed for larger more controllable thermal generating plants. Allowing the aggregation of DERs to participate in markets is a step closer to promoting DER development.

North American grid operators can also draw upon the approaches taken by European electricity markets as recommended by the Smart Energy Demand Coalition (SEDC) [30, 641]. The SEDC noted that favorable regulation and market rules, in addition to promoting DR programs, were key to the successful integration of variable energy resources in the European electric power industry [30, 641].

North American utilities have a chance to take on the additional roles created by DERs to maximize their returns as well as ease the integration of DERs. Traditionally, the interaction between utilities and consumers has been limited to maintaining the distribution service, responding to the occasional call whenever supply is interrupted and providing metering/billing services [30]. However, as more DERs are installed on the distribution system, utilities have the chance to expand their services beyond network upgrades and potentially assume the role of a DSO and control services such as DR and curtailment. Furthermore, DERs offer many flexibilities that could be leveraged by utilities to reduce system and operational costs [30]. For example, an increase in distributed solar PV systems could result in operational challenges that could be mitigated by enabling inverter control to regulate both the quality and quantity of PV power sent to the distribution feeders [30]. Additionally, distributed energy storage could support solar PV production, thus significantly reducing the need for system and network upgrades [30].

However, it is important to note that at current battery costs, network upgrades might be more affordable compared to installing new energy storage infrastructure. As for assuming the role of DSOs, favorable regulation is necessary to ensure a level playing field for all DERs and enable any new stakeholders [30, 69]. A revision of market rules to allow DERs to participate in markets competitively would be necessary as well as ensuring transparency in the ownership and control of DER operations [30].

Of course, the effective control of DERs requires strictly laid out guidelines on the eligibility, metering, telemetry, and operational coordination between RTO/ISO's, DER aggregators, and distribution utilities [609]. It is likely that new stakeholders will step up and assume the role of controlling and easing the integration of DERs. At the moment, however, distribution utilities are well placed to undertake these additional responsibilities given their awareness of both generation and the consumption flexibility of consumers and DERs [642].

Proper management of DERs and TE frameworks would result in a dynamic distribution system that is centered on energy products, regulation products, and time-responsive prices that help stabilize the grid through the provision of energy balancing, line congestion management, and voltage control [30, 643]. As in the case of European power markets, utilities may need to assume the role of the DSO. This would constitute a tremendous change in the utility business models and current regulatory structures [30].

The question of whether utilities need to be deregulated to allow for this transition must also be considered. For a long time, utilities have enjoyed a natural monopoly status that needs to be unbundled to allow for competition in the markets and encourage the presence of DER aggregators at the distribution level [643]. Assuming utilities take on the role of a DSO, their relationship with consumers must transform into a partnership where the utilities, such as DSOs, engage with prosumers to achieve the common goal of the partial supply of services [643]. This symbiotic relationship between consumers and utilities is best summarized in Fig. 4.4, where a smart home with several DERs interfaces with the grid to provide and receive services as necessary. As a DSO, a utility can serve as an intermediary

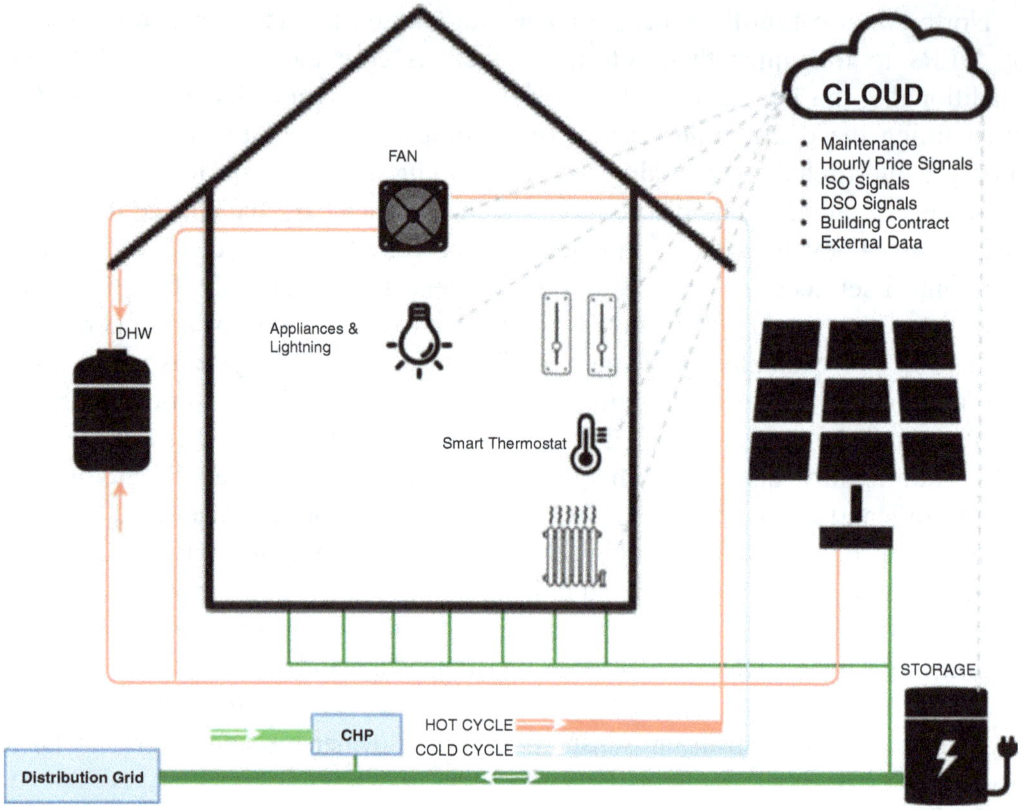

Fig. 4.4 An example eIoT-enabled smart home: DERs are connected to the grid through a cloud-based framework (adapted from [30])

to balance the supply and demand of power while correcting for any surpluses and stability issues quickly and reliably [643].

The transformation of the grid is already underway and it puts pressure on utilities to adapt to the competition and become an integral part of the future grid. Competition at the distribution level is set to increase with the presence of DR aggregators and peer-to-peer electricity trading platforms [644]. Although the distribution system has not been as observable as the transmission system, smart meters and remote terminal units (RTUs) are quickly closing this gap [107, 645]. As a result, the role of utilities is set to transform to a more active one that is very similar to the role of transmission system operators (TSO) [646, 647]. Utilities, such as DSOs, would potentially serve as neutral market facilitators to guarantee system stability and power quality while ensuring technical efficiency and fair prices for all parties involved [646].

The adoption of eIoT and TE management platforms for grid monitoring and control will result in large quantities of data that requires management [645, 648]. Needless to say, neutrality, transparency, and non-discriminatory data management are highly necessary to ensure a level playing field for all market participants [648]. The European Union serves as a great example for the creation of DSOs and the adoption of eIoT. Organizations such as the SEDC [641] and EURELECTRIC have played key roles in identifying the potential challenges of integrating variable energy resources and the eIoT. Many of these lessons have applications to the North American electric grid. The strategic direction and role of electric utilities in this new landscape remains unclear and depends on the answers to several open questions:

1. Which agent will evaluate and deploy aggregated DERs? The utility? The aggregator? The RTO/ISO?
2. Which entity will manage and prioritize DER dispatch?
3. How will stakeholders address concerns about possible double compensation?
4. What level of visibility will distribution utilities and RTOs/ISOs need into the operations of aggregated DERs to reliably manage those assets?
5. Which entity pays for distribution system upgrades needed to facilitate DER participation in wholesale markets?
6. How will utilities recover costs to enable DER aggregation within their territories?
7. How will the evolving technological landscape of eIoT affect the answers to these questions?
8. How will FERC-level regulations affect the answers to these questions?

4.4 Customer Applications

4.4.1 Industrial Applications

The industrial sector consumes approximately 42% of all the electricity produced in the world [649]. Apart from being energy intensive, some manufacturing processes, such as with electrical drives and motors, demand high-quality electricity [650]. In addition, the industrial sector is facing high pressure to decarbonize from both regulation [651, 652] and corporate social responsibility [653, 654]. As a result, most industrial facilities have integrated on-site DERs and are rapidly undertaking energy efficiency measures to minimize their carbon footprint [655].

In most cases, the energy requirements of industrial facilities cannot be served by only a local utility. Hence, these facilities sometimes directly connect to the transmission lines and participate in the wholesale electricity markets. Typically, industrial electric loads are consistent, large scale, and centralized [649], making them good candidates for DR programs. In some countries, industrial base loads have been used by system operators for the provision of various ancillary services [649]. As it happens, it is much easier to control a few large industrial loads than numerous small residential loads. Furthermore, recognizing the higher (economic) utility of consumed electricity for industrial processes, it can be expected that production systems will be more willing to respond to price signals in DSM schemes to ensure steady and continuous supply.

The nature of industrial loads provides an easy opportunity to apply DSM to industrial energy systems [649]. The ability to reschedule or "shift" loads is particularly important as more solar and wind resources are added to the grid. At present, DSM applications compensate consumers based on their load reduction from a predefined baseline. However, studies have shown that the process of determining the baseline is prone to errors likely to cost more and result in other system imbalances that could propagate through various layers of power system control [250, 656, 657]. The industrial sector, however, provides many opportunities for load shifting that if scheduled and coordinated properly could improve DSM applications. Not only does load shifting increase demand flexibility, it also ensures that power quality is maintained [649]. That said, industrial processes that are not time constrained can be scheduled so that they can shift demand to help balance the electricity grid under certain demand constraints.

In the same way, constrained industrial processes could store intermediate power for use during periods of high demand. Currently, storage is being used in industry in the form of pumped hydro, compressed air, hydrogen, batteries, flywheels, superconducting magnetic energy storage, and super-capacitors [649] to support various applications. While storage increases flexibility, there is a decrease in efficiency, since transferring electricity to and from storage devices is not 100% efficient [649].

The concept of IoT is not new to industrial applications. IoT has been supporting industrial and manufacturing processes for over a decade now, with applications in business continuity management, anomaly detection as well as supply-chain management [658]. These IoT applications provide a control platform that could be used to carry out various DR functions. Obviously, equipment upgrades may be necessary to provide the connection and coordination capabilities for eIoT devices.

As discussed in Sect. 3.1.5, the main barrier to the adoption of eIoT lies in the cost of sensors, especially for small-scale consumers of electricity. However, industrial consumers are able to diffuse the energy cost management across various layers due to economies of scale for the required improvements. Additionally, most industries already monitor load data in real time and possess the necessary smart metering and data exchange equipment that will eventually reduce the investment cost in eIoT infrastructure [649]. These factors significantly simplify the adoption and application of eIoT in industrial energy-management applications. In fact, this makes the industrial consumer well suited for the use case discussed in Sect. 4.2.2. As stated in Sect. 3.2.4.6, IIoT and eIoT devices are overlapping and complementary rather than mutually exclusive. Therefore, the development of eIoT within industrial applications will go hand in hand with the current IIoT implementations.

4.4.2 Commercial Applications

The majority of electricity consumed in the USA goes to commercial and residential building energy systems [607]. According to the US Energy Information Administration (EIA), 77.46% of electricity generated in January 2018 was consumed by commercial and residential buildings [659]. Traditionally, commercial buildings have included hospitals, hotels, stores, and offices [660]. Commercial buildings come in a variety of sizes, and depending on the services the business provides, are less flexible to participate in DSM programs. For example, a hospital requires access to energy 24/7 and would be less willing to participate in an interruptible program [660].

In recent times, decabornization and sustainability concerns have driven most commercial enterprises to seek cleaner alternative sources of energy such as wind and solar. For some, this sustainable transition has been composed of a mix of energy efficiency measures and investment in renewable energy resources. Companies with large servers have shown great commitment to decarbonizing with some like Google vowing to source 100% of their energy from renewable sources by 2017 [661, 662]. As signatories of the Department of Energy's Better Buildings Initiative, various commercial corporations such as Walmart have committed to reduce over 20% of their energy consumption and as of 2018 they sourced approximately 28% of their total electricity from renewable sources [663].

eIoT is going to play a key role in ensuring grid reliability especially as more and more commercial enterprises assume the role of prosumers. In time, commercial enterprises such as Google and Walmart will become energy independent. Naturally, this implies more flexibility and freedom to directly participate in electricity wholesale markets. Without demand-side options that offer the equivalent (if not better) rewards for these corporations to trade and manage their energy, commercial enterprises will most certainly bypass utilities altogether. TE applications have an active role to play in creating platforms that engage commercial consumers at this level of the electric grid value chain. Most commercial buildings possess various eIoT capabilities in energy load management applications such as HVAC, and lighting [664]. For some commercial consumers such as grocery stores, sophisticated dynamic energy-management capabilities are necessary to maintain steady operation of their facilities. For example, department stores would prefer a positive pressure differential so that the air leakage happens outward instead of inward.

The implementation of eIoT for commercial customers will take many shapes depending on the services and type of the commercial entity. However, certain energy-management solutions such as smart metering, and price incentives could be used to advance the energy supply and control for these consumers. Net metering is expected to become a common practice in both commercial and residential buildings that want to be incorporated into utility planning and price structures. So far, 43 out of 50 US states have established net metering policies to support such engagements [665].

Unlike residential buildings, commercial building owners have a fixed decision-making structure that is most ideal for participation in demand-side programs. Usually, owners of commercial buildings are more sensitive to price incentives and most commonly have a single owner to expedite decision-making. Price incentives have encouraged the adoption of smart building management systems, where buildings are actively managing energy consumption. This means that building owners may soon become participants in real-time energy markets [666]. Requirements for this future development in energy management include automatic operational control capabilities for building subsystems, such as HVAC and lighting, and real-time communication with the grid [666].

Whether implementing DSM or individually engaging in energy pricing arbitrage, a variety of data coordination with system operators or utilities is necessary. Third parties such as energy aggregators and energy service companies are expected to use eIoT to improve energy-conservation savings [667]. This can be achieved through the installation of sensors that can monitor progress, and platforms for building management systems [668].

Recent studies have predicted a steady growth in the deployment of building energy-management systems (BEMS) for commercial as well as residential buildings. BEMS have attracted a lot of funding (more specifically $1.4B between 2000–2014) and are set to revolutionize the operations and control of commercial and residential buildings [669]. The US Department of Energy estimates that by 2020, BEMS applications will comprise 77% of the $2.14 billion US market [670, 671]. This implies an increase in sensors and internet-connected devices to manage and control building energy consumption.

Internet connectivity results in security concerns that are hopefully addressed by having cloud-hosted BEMS to relieve consumers of the need to secure their own devices or web-enabled services [672]. With time, the overall awareness and control for operators, consumers and owners will significantly improve and thus simplify the integration of renewable energy resources, energy storage, and electric vehicles. BEMS provide a key opportunity for TE-based frameworks to control, coordinate, and negotiate transactions among connected devices. For commercial customers, eIoT could be leveraged to reduce the overall energy consumption as well as improve the operation of these energy-intensive systems. As more commercial consumers adopt eIoT, they will be well placed to employ either of the two use cases described in Sect. 4.2.

4.4.3 Residential Applications

Another key TE application area is in the energy-management solutions for residential customers. Unlike commercial and industrial customers, residential consumers consume smaller loads and their energy decisions are very much comfort driven. In addition, heterogeneity in home infrastructures poses difficulties in smart energy management, since communication is required between the system, customer users, energy devices, and system operators [673]. Given the high cost of sensors, most residential customers may be reluctant to adopt new and improved sensors.

That said, the overall public opinion is shifting towards cleaner and more sustainable energy solutions. A significant percentage of the population is either producing their own electricity or opting to purchase only renewable energy.

As more residents become prosumers and sustainable, an increase in residential microgrids is expected. Naturally, TE platforms could assume the role of negotiating transactions for such microgrids as addressed in the first use case or through direct participation in wholesale energy markets in the second use case.

TE platforms for residential customers must provide an enhanced user experience and incentives that influence consumer behaviors. Consumer behaviors can be influenced through techniques such as real-time consumption monitoring, ubiquitous sensing, or contextual comparisons with neighbors [673]. However, this ubiquitous influence raises privacy and security concerns, which need to be carefully addressed especially if the data collected is to be used to gather insight on consumer behavior, build intelligent modeling tools, and support automatic grid operations [673].

As the number of smart devices in the home rises, platforms that allow interoperability among smart devices and provide a hub for consumers to customize their devices are necessary. So far, consumer apps such as Stringify and If This, Then That (IFTTT) offer options to connect similarly used devices and to create conditional statements for controlling remote devices, respectively. A key device in a residential home that is easily controlled through such applications is the smart thermostat.

As of September 2017, there were over two million smart thermostats, and a recent Navigant report predicts a four million rise by 2024 [608]. Several models have emerged for the control integration of smart thermostats including through utilities, by self-install, or in Bring Your Own Thermostat (BYOT) programs [608]. Another approach is the direct control of thermostats, which currently has an opt-out rate of 21% [608]. High opt-out rates as well as recruiting new customers, maintaining old customers, and device interoperability are key challenges [608] that still face the implementation of TE-based platforms in residential homes.

Given the high preference for comfort, privacy, and convenience, a single platform for DR and device control would work best for residential homes. Currently, utilities lack a single, all-encompassing program for DSM. About 16% of utilities offer water heater programs and 24% offer thermostat control programs, while only 9% provide behavioral programs to their residential customers [608]. However, due to reliability concerns, only half of these programs were actually called upon to provide DR in 2016 [608]. In addition, a wide range of DR options is necessary to enable more consumers to participate. As these programs evolve with real-time eIoT, DSM programs must shift from their current annual load shaping perspectives to less-than-a-minute perspectives for the provision of ancillary services. "Shape, Shift, Shed, Shimmy" is a framework built in California that incorporates timescales to better understand how to use DR.

Electric load from electric vehicles (EVs) is set to significantly increase residential loads requiring a framework to manage and control the power consumption of EVs. The power consumed by EVs is expected to reach 400 TWh annually by 2040 [608]. TE DR platforms for EVs are essential to manage this disruptive technology. Studies have shown that EVs could be used as flexible loads for the provision of ancillary services if managed properly. Currently, 19% of utilities are offering EV DR programs, while 79% are either planning or researching the DR potential of EVs [608].

Managed charging, either through utilities, load-balancing authorities, or aggregators, allows EVs to be used as storage to absorb excess renewable energy generation and smooth adverse effects on the net load [79–88]. From a technology point of view, TE platforms will require investments in infrastructure to support communication signals sent between a vehicle, other vehicles, home systems, and grid operators. Although behavioral programs could be used to affect charging times or quantity, technical integration is necessary to extract other potential grid service values in capacity, emergency load reduction, reserves, and renewable energy absorption [608]. All in all, electric vehicles offer great potential for DR that could be leveraged in a number of ways to support grid operations.

Residential TE applications stand to benefit from using behavioral DR to curb peaks, increase consumer participation and savings, and reduce the cost of engaging the large residential consumer base. Although implementation of these applications still faces many challenges, optimizing how a customer is contacted, determining how far in advance to notify a customer of an event, communicating why an event is called, how the program works and how a customer can participate, and strategically planning event calls will go a long way to ensure customer retention. Due to its analytical benefits, eIoT is likely to be instrumental in deploying behavioral demand response programs.

Chapter 5
eIoT Transforms the Future Electric Grid

In conclusion, the development of eIoT is an integral part of the transformation to the future electricity grid.

- Chapter 1 discussed five energy-management change drivers that are bringing about this transformation:
  - Rising demand for electricity [34–36]
  - Emergence of renewable energy resources [37–40]
  - Emergence of electrified transportation [41, 42]
  - Deregulation of power markets [43, 44]
  - Innovations in smart grid technology [45, 46]
- Chapter 2 explained that the impact of these energy-management change drivers will appear primarily at the grid's periphery. Distributed generation in the form of solar photovoltaics (PV) and small-scale wind will be joined by a plethora of internet-enabled appliances and devices to transform the grid's periphery to one with two-way flows of power and information [45, 46]. The resulting activation of the grid periphery gives rise to an energy internet of things composed of network-enabled physical devices, heterogeneous communication networks, and distributed control and decision-making algorithms.
- Chapter 3 organized the discussion of these elements into an eIoT control loop built upon well-established standards and architectures.
- Chapter 4 showed that such an eIoT control loop is most consonant with the emerging concept of TE and then proceeded to discuss how it may be applied within utilities–distribution system operators and industrial, commercial, and residential customers.

In summary, eIoT is set to transform all aspects of grid operations and control. This transformation spans both technical and economic layers and leads to new applications, stakeholders, and energy system management solutions. This chapter serves to summarize the conclusions of the work: 1. eIoT will become ubiquitous, 2. eIoT will enable new automated energy-management platforms, and 3. eIoT will

S. O. Muhanji et al., *eIoT*, https://doi.org/10.1007/978-3-030-10427-6_5

enable distributed techno-economic decision making. This chapter also serves to highlight two open challenges and opportunities for future work: the convergence of cyber, physical, and economic performance, and the re-envisioning of the strategic business model for the utility of the future. These conclusions and open challenges are now discussed in turn.

5.1 Conclusions

5.1.1 eIoT Will Become Ubiquitous

As discussed previously in Sect. 3, the number of sensors and actuators deployed at all levels of the electric grid is set to dramatically increase. These sensors and actuators will enable the transformation of both the distribution and transmission network aiding in the measurement and actuation of primary and secondary electric power variables. The transformation is going to be characterized by improvement in the quality of data measured and a significant increase in the diversity of measurements taken. The speed and granularity of measured variables in the transmission system will be enhanced through widespread adoption of PMUs, and an upgrade of the SCADA system as addressed in Sect. 3.1.2.1. Monitoring of secondary variables such as wind speed and solar irradiance will significantly improve the forecasting accuracy and capability, and promote the overall reliability of the supply of wind and solar power.

The steady supply of natural gas is critical to ensuring electric power supply reliability especially with major base load retirements. This motivates the need for secondary measurements by eIoT to ensure reliable and cost-effective operation of the electric and natural gas supply systems as covered in Sect. 3.1.3. As for transmission system actuation, the adoption of decentralized or distributed approaches for AGC and AVR applications is imperative to effectively control distributed energy resources. Naturally, current FACTS devices must also become smarter to enable faster, efficient, and accurate measurement and actuation of transmission variables as discussed in Sect. 3.1.2.2.

Advanced metering infrastructure with AMR and AMM capability provides access to consumer data and enables two-way communication between consumer devices and utilities. Smart sensing devices will also motivate consumers to upgrade their homes for faster and efficient energy management. Energy monitors, smart switches, outlets, lights, and HVAC will provide better actuation abilities for consumers while allowing for secondary measurements that would ultimately improve the efficiency of DR programs.

The mere presence of sensors and actuation devices triggers innovations and advancements in the communication networks that connect them. Communication such as SCADA networks and wide area monitoring systems are expected to continue to play an integral role in utility and grid operator communication networks.

Low power wide area networks will allow communication over long ranges while minimizing the energy consumption of devices. Communication devices that go beyond the purview of either utility or grid operators will be needed to enable the inclusion of all interacting parties. Telecommunication networks may need to take on the role previously carried out by utility and grid networks. Local area networks will play a key role in ensuring the full automation of residential, industrial, and commercial premises. Together, these networks will create a web of interacting devices that will work collaboratively to ensure the reliability of the electric power supply system. Furthermore, this network of interacting devices will enable the emergence of TE platforms that will revolutionize the exchange of energy products and services.

5.1.2 eIoT Will Enable New Automated Energy-Management Platforms

eIoT will create a network of interacting devices that measure, store, and actuate data in real-time. These devices also bring about many opportunities for the improvement of current electric power system operations. Most of these opportunities are observed at the grid periphery where millions or even billions of interacting devices will emerge in turn to create numerous control points for the distribution grid. The once passive consumer base will become active participants in their own energy supply and consumption. While some consumers will become prosumers, others will have the opportunity to participate in electricity markets or carry out transactions with their neighbors. In addition, the grid periphery will be characterized by a proliferation of DERs such as rooftop solar and electric vehicles that need management.

The transformation of the grid periphery calls for several changes to status quo. The distribution network will require an upgrade and depending on the issue, non-wire solutions such as engaging consumers through DR may be necessary. This calls for better energy-management platforms that help engage the consumer base. As DERs begin to participate directly in electricity markets, aggregation platforms or companies will be necessary to avoid any reliability issues. A change in the regulatory or market structure may be required to aid in the smooth participation of DERs and efficient DR programs.

Depending on the willingness of utilities to step up to these new challenges, this could result in the transformation of the utilities business model or the emergence of new stakeholders to take on these new roles. Either way, the effective deployment of eIoT will require new energy-management platforms whether they are for managing energy transactions or for managing the large quantities of data collected in real-time. TE and blockchain-powered platforms are starting to emerge as potential energy-management platforms. Additionally, various cloud-based commercial IoT platforms such Amazon, Microsoft, SAP, and OpenFog are emerging to support the millions of interacting IoT devices. With time, these platforms will also evolve to specifically cater to the energy industry.

5.1.3 eIoT Will Enable Distributed Techno-Economic Decision Making

In order to control the millions or even billions of interacting devices, scalable and distributed techno-economic decision making will be needed. Whether it is in the transmission system with distributed AGC and AVR or in the control of smart devices at the grid periphery, distributed control will play a key role in the effective deployment of eIoT devices. Through TE, eIoT will enable distributed decision making of physical and economic power supply variables. The eIoT control loop is centered around sensing, communication, actuation, and distributed control algorithms that creates an effective decision-making framework. This framework informs and executes complex decisions that are spawned by distributed technical and economic information from all over the electric power supply and distribution system. The distributed economic decision making will greatly benefit DR applications through peer-to-peer trading platforms and smart energy-management programs.

5.2 Challenges and Opportunities

5.2.1 The Convergence of Cyber, Physical, and Economic Performance

eIoT is not without its challenges. With every challenge, comes an opportunity to advance the electric power system. eIoT causes a convergences of the cyber, physical, and economic performance of the electricity grid.

- Most eIoT devices will have and/or require an internet connection.
- eIoT devices need to work together to perform different functions across the electric supply and demand value chain.
- New market participants such as aggregators, prosumers, DERs, and microgrids will emerge.
- A large quantity of data will be generated and stored or processed in real-time.

Connecting eIoT devices to the internet creates a cybersecurity concern for grid operators and all parties involved. This requires investment in technologies to ensure the integrity and security of all devices in the network.

Additionally, careful vetting of interacting devices may be necessary to prevent infections from spreading through rogue devices or connections. Data sent to the cloud must also be vetted to avoid exposing sensitive data to security issues. This may require equipping devices with enough processing capabilities to carry out some computations without involving the cloud. The electric grid architecture is increasingly transforming, more specifically, to one with two-way flows of power and information. This architecture creates a cyber–physical requirement where both physical devices and informatic components must accommodate this architectural need. With changing architectural requirements, the cyber–physical–economic aspects of the grid must be designed in such a way as to ensure interoperability. This provides an opportunity for the development of standards for ensuring interoperability.

The emergence of new market participants creates the need for more devices, platforms, and economic structures not to mention regulatory changes to manage and control their participation in electricity markets.

A mechanism to store, manage, and secure the data collected in real-time is necessary to protect the interests of all stakeholders. Although the convergence of the cyber, physical, and economic aspects of grid operations poses a challenge, it provides an opportunity for collaboration across various layers of the electricity grid and jurisdictions to enhance system reliability.

5.2.2 *Re-envisioning the Strategic Business Model for the Utility of the Future*

The biggest transformation will occur on the distribution side at the grid periphery. In addition to the millions of interacting devices, the rise in the number of active consumers and DERs poses a major challenge to the utility business model. Utilities must re-evaluate their approach to how they manage their system. For example, instead of defaulting to network upgrades to accommodate DERs, utilities may consider the potential of non-wire solutions.

In order to engage the active consumer base, utilities must develop proper compensation mechanisms that:

1. Motivate consumers to shift and/or lower consumption
2. Are fair and offer value to the consumer
3. Provide a diversity of options that cater to varying consumer needs.

This may require either a complete transformation of the utility business model or open collaboration with aggregators and emerging stakeholders. The distribution market structure may transform to be similar to that of the wholesale electricity markets observed at the ISO/RTO level. This, in turn, may require regulatory measures that foster fair and competitive markets to equally engage all participants.

The deployment of eIoT poses numerous challenges that span the cyber, physical, economic, and regulatory structure of the electricity supply and demand value chain. A holistic approach is necessary to effectively deal with these challenges. Consequently, stakeholders at various jurisdictional layers must engage with each other to work out a favorable solution that benefits most if not all. The success of this collaboration highly depends on the existence of favorable regulatory and policy structures as well as standards that serve as guidelines for stakeholders.

References

1. EIA, 2016 Annual Energy Outlook. United States Energy Information Administration, Tech. Rep., 2016
2. LLNL, Estimated U.S. Energy Consumption in 2015 Energy Flow Chart. Lawrence Berkeley National Laboratory, Livermore, CA, Tech. Rep., 2016. https://flowcharts.llnl.gov/content/assets/docs/2015_United-States_Energy.pdf
3. A. Ipakchi, F. Albuyeh, Grid of the future. IEEE Power Energ. Mag. **7**(2), 52–62 (2009)
4. California Independent System Operator (CAISO), What the Duck Curve Tells us about Managing a Green Grid. California Independent System Operator (CAISO), Tech. Rep., 2016. https://www.caiso.com/Documents/FlexibleResourcesHelpRenewables_FastFacts.pdf
5. D.C. Mcfarlane, S. Bussmann, Developments in holonic production planning and control. Prod. Plan. Control **11**(6), 522–536 (2000)
6. D. Mcfarlane, Auto-ID based control: an overview. University of Cambridge–Institute for Manufacturing AUTO-ID Centre, Tech. Rep. CAM-AUTOID-WH-004 (2002)
7. D. McFarlane, M. Kollingbaum, J. Matson, P. Valckenaers, Development of Algorithms for agent-based control of manufacturing flow shops. Stud. Inf. Control **11**(1), 41–52 (2002)
8. D. Mcfarlane, V. Agarwal, A.A. Zaharudin, C.Y. Wong, R. Koh, Y. Kang, The intelligent product driven supply chain, in *Proceedings of IEEE International Conference on Systems, Man and Cybernetics*, vol. 4 (IEEE, Piscataway, 2002), pp. 393–398
9. D. McFarlane, Y. Sheffi, Mcfarlane, sheffi–2003–the impact of automatic identification on supply chain operations.pdf (2003)
10. D. McFarlane, S. Sarma, J.L. Chirn, K. Ashton, C.Y. Wong, Auto ID systems and intelligent manufacturing control. Eng. Appl. Artif. Intell. **16**(4), 365–376 (2003)
11. D. McFarlane, S. Bussmann, S.M. Deen, Holonic manufacturing control: rationales, developments and open issues, in *Agent-Based Manufacturing* (Springer, Berlin, 2003), pp. 303–326
12. J. Jarvis, D. Jarvis, D. McFarlane, Achieving holonic control-an incremental approach. Comput. Ind. **51**(2), 211–223 (2003)
13. N. Chokshi, A. Thorne, D. Mcfarlane, White paper routes for integrating auto-ID systems into manufacturing control middleware environments. Auto-ID Centre (2004), pp. 0–27
14. F.A. Rahimi, A. Ipakchi, Transactive energy techniques: closing the gap between wholesale and retail markets. Electr. J. **25**(8), 29–35 (2012)
15. A.M. Farid, B. Jiang, A. Muzhikyan, K. Youcef-Toumi, The need for holistic enterprise control assessment methods for the future electricity grid. Renew. Sust. Energ. Rev. **56**(1), 669–685 (2015). http://dx.doi.org/10.1016/j.rser.2015.11.007

S. O. Muhanji et al., *eIoT*, https://doi.org/10.1007/978-3-030-10427-6

16. S.O. Muhanji, A. Muzhikyan, A.M. Farid, Long-term challenges for future electricity markets with distributed energy resources, in *Smart Grid Control: An Overview and Research Opportunities*, ed. by J. Stoustrup, A.M. Annaswamy, A. Chakrabortty, Z. Qu (Springer, Berlin, 2017)
17. S.O. Muhanji, A. Muzhikyan, A.M. Farid, Distributed control for distributed energy resources: long-term challenges and lessons learned. IEEE Access **1**(1), 1–17 (2018)
18. A. Phillips, L. van der Zel, *Sensor Technologies for a Smart Transmission System* (Electric Power Research Institute, Palo Alto, 2009)
19. S.A. Boyer, *SCADA- Supervisory Control And Data Acquisition*, 3rd edn. (ISA, Research Triangle Park, 2004)
20. A. Gomez-Exposito, A.J. Conejo, C. Canizares, *Electric Energy Systems: Analysis and Operation* (CRC Press, Boca Raton, 2008)
21. O.O.E.D., U.S. Department of Energy and E. Reliability, Distribution automation: results from the smart grid investment grant program. U.S. Department of Energy, Office of Electricity Delivery and Energy Reliability, Tech. Rep., 2016. https://www.energy.gov/sites/prod/files/2016/11/f34/Distribution%20Automation%20Summary%20Report_09-29-16.pdf
22. The Smart Grid Interoperability Panel–Cyber Security Working Group and W. Group, Guidelines for smart grid cyber security: vol. 2, privacy and the smart grid, in *National Institue of Standards and Technology US Department of Commerce*, vol. 2 (National Institute of Standards and Technology, U.S. Department of Commerce, Washington, 2010), pp. 1–69
23. Office of Energy Efficiency and Renewable Energy (EERE-DOE), Manufacturing Energy and Carbon Footprints (2010 MECS) (2010). https://www.energy.gov/eere/amo/manufacturing-energy-and-carbon-footprints-2010-mecs
24. P. Jadun, C. McMillan, D. Steinberg, M. Muratori, L. Vimmerstedt, T. Mai, Electrification Futures Study: End-Use Electric Technology Cost and Performance Projections Through 2050. NREL, Tech. Rep., 2017
25. E. Fadel, V.C. Gungor, L. Nassef, N. Akkari, M.A. Malik, S. Almasri, I.F. Akyildiz, A survey on wireless sensor networks for smart grid. Comput. Commun. **71**, 22–33 (2015)
26. M. Gottschalk, M. Uslar, C. Delfs, *The Use Case and Smart Grid Architecture Model Approach: The IEC 62559-2 Use Case Template and the SGAM Applied in Various Domains* (Springer, Berlin, 2017)
27. GWAC, Smart Grid Interoperability Maturity Model Beta Version. Gridwise Architecture Council, Tech. Rep., 2011
28. R. Melton, Gridwise Transactive Energy Framework Version 1. Grid-714 Wise Archit. Council, Richland, Tech. Rep. PNNL-22946, vol. 715, p. 716 (2015)
29. International Electrotechnical Commission, IEC TR 62357-1:2016 Power Systems Management and Associated Information Exchange. Part 1: Reference Architecture (2016)
30. I. Pérez-Arriaga, C. Knittle, *Utility of the Future: An MIT Energy Initiative Response to an Industry in Transition* (MIT Energy Initiative, Cambridge, 2016)
31. A.M. Farid, Electrified transportation system performance: conventional vs. online electric vehicles, in *The On-line Electric Vehicle: Wireless Electric Ground Transportation Systems*, ed. by N.P. Suh, D.H. Cho, chap. 20 (Springer, Berlin, 2017), pp. 279–313. http://engineering.dartmouth.edu/liines/resources/Books/TES-BC05.pdf
32. A.M. Farid, Multi-agent system design principles for resilient coordination and control of future power systems. Intell. Ind. Syst. **1**(3), 255–269 (2015). http://dx.doi.org/10.1007/s40903-015-0013-x
33. P. Schavemaker, L. Van der Sluis, Books24×7 Inc., in *Electrical Power System Essentials* (Wiley, Chichester, 2008). http://www.loc.gov/catdir/enhancements/fy0810/2008007359-d.html; http://www.loc.gov/catdir/enhancements/fy0810/2008007359-t.html
34. IEA, World Energy Outlook, Energy and Air Pollution. International Energy Agency, Paris, Tech. Rep., 2016
35. F. Birol et al., World Energy Outlook 2010. International Energy Agency, vol. 1 (2010)
36. F. Birol, E. Mwangi, U. Global, D. Miller, S. Renewables, P. Horsman, Power to the people. IAEA Bull. **46**, 9–12 (2004)

37. E. Commission, *A Roadmap for Moving to a Competitive Low Carbon Economy in 2050* (European Commission, Brussel, 2011)
38. M. Haller, S. Ludig, N. Bauer, Decarbonization scenarios for the EU and MENA power system: considering spatial distribution and short term dynamics of renewable generationnamics of renewable generation. Energy Policy **47**, 282–290 (2012)
39. J. Rogelj, M. Den Elzen, N. Höhne, T. Fransen, H. Fekete, H. Winkler, R. Schaeffer, F. Sha, K. Riahi, M. Meinshausen, Paris Agreement climate proposals need a boost to keep warming well below 2 C. Nature **534**(7609), 631–639 (2016)
40. W. Obergassel, C. Arens, L. Hermwille, N. Kreibich, F. Mersmann, H.E. Ott, H. Wang-Helmreich, Phoenix from the ashes—an analysis of the Paris agreement to the United Nations framework convention on climate change. Wuppertal Inst. Clim. Environ. Energy **1**, 1–54 (2016)
41. G. Pasaoglu, M. Honselaar, C. Thiel, Potential vehicle fleet CO_2 reductions and cost implications for various vehicle technology deployment scenarios in Europe. Energy Policy **40**, 404–421 (2012)
42. T. Litman, *Comprehensive Evaluation of Transport Energy Conservation and Emission Reduction Policies* (Victoria Transport Policy Institute, Victoria, 2012)
43. S. Stoft, *Power System Economics: Designing Markets for Electricity*, 1st edn. (Wiley-IEEE Press, Hoboken, 2002)
44. W.W. Hogan, Electricity whole-sale market design in a low-carbon future, in *Harnessing renewable Energy in Electric Power Systems: Theory, Practice, Policy* (RFF Press, Washington, 2010), pp. 113–136
45. P. Palensky, D. Dietrich, Demand side management: demand response, intelligent energy systems, and smart loads. IEEE Trans. Ind. Inf. **7**(3), 381–388 (2011)
46. P. Siano, Demand response and smart grids—a survey. Renew. Sustain. Energy Rev. **30**, 461–478 (2014)
47. E.G. Cazalet, Automated transactive energy (TeMIX), in *Grid-Interop Forum* (TeMIX, Los Altos, 2011)
48. IEA, World Energy Outlook 2016. Part B: Special Focus on Renewable Energy. International Energy Agency, Tech. Rep., 2016
49. J.H. Williams, B. Haley, F. Kahrl, J. Moore, A.D. Jones, M.S. Torn, H. McJeon et al., Pathways to deep decarbonization in the united states, in *The US Report of the Deep Decarbonization Pathways Project of the Sustainable Development Solutions Network and the Institute for Sustainable Development and International Relations*, vol. 23 (Energy and Environmental Economics, San Francisco, 2014), p. 2016
50. J. Williams, Policy implications of deep decarbonization in the United States, in *AGU Fall Meeting Abstracts* (2015)
51. J.K. Kaldellis, D. Zafirakis, The wind energy (r) evolution: a short review of a long history. Renew. Energy **36**(7), 1887–1901 (2011)
52. L.H. Hansen, P.H. Madsen, F. Blaabjerg, H. Christensen, U. Lindhard, K. Eskildsen, Generators and power electronics technology for wind turbines, in *IECON'01. 27th Annual Conference of the IEEE Industrial Electronics Society*, vol. 3 (IEEE, Piscataway, 2001), pp. 2000–2005
53. Y. Kumar, J. Ringenberg, S.S. Depuru, V.K. Devabhaktuni, J.W. Lee, E. Nikolaidis, B. Andersen, A. Afjeh, Wind energy: trends and enabling technologies. Renew. Sustain. Energy Rev. **53**, 209–224 (2016)
54. IEA, Renewables 2017 Analysis and Forecasts to 2022. International Energy Agency, Tech. Rep., 2017
55. N. Kannan, D. Vakeesan, Solar energy for future world—a review. Renew. Sustain. Energy Rev. **62**, 1092–1105 (2016)
56. C.A.T. Partners, Climate Action Tracker: China (2018). http://climateactiontracker.org/countries/china.html
57. Z.-Y. Zhao, J. Zuo, L.-L. Fan, G. Zillante, Impacts of renewable energyregulations on the structure of power generation in China—a critical analysis. Renew. Energy **36**(1), 24–30 (2011)

58. D. Zhang, J. Wang, Y. Lin, Y. Si, C. Huang, J. Yang, B. Huang, W. Li, Present situation and future prospect of renewable energy in China. Renew. Sustain. Energy Rev. **76**, 865–871 (2017)
59. Z. Ming, L. Ximei, L. Na, X. Song, Overall review of renewable energy tariff policy in China: evolution, implementation, problems and countermeasures. Renew. Sustain. Energy Rev. **25**, 260–271 (2013)
60. Z. Ming, L. Ximei, L. Yulong, P. Lilin, Review of renewable energy investment and financing in China: status, mode, issues and countermeasures. Renew. Sustain. Energy Rev. **31**, 23–37 (2014)
61. US Department of Energy USDOE, Business energy investment tax credit (ITC) (2010). https://energy.gov/savings/business-energy-investment-tax-credit-itc
62. N. Groom, Republican tax bill hits wind power, solar largely unscathed (2017). https://www.reuters.com/article/us-usa-tax-renewables/republican-tax-bill-hits-wind-power-solar-largely-unscathed-idUSKBN1D22R9
63. USIRS, Renewable Electricity Production Tax Credit (PTC) (2016). https://www.energy.gov/savings/renewable-electricity-production-tax-credit-ptc
64. USDOE, Advancing the Growth of the U.S. Wind Industry: Federal Incentives, Funding, and Partnership Opportunities. U.S. Department of Energy, Tech. Rep., 2017. https://www.energy.gov/sites/prod/files/2017/02/f34/67742_0.pdf
65. USDOE, Alternative Fuels Data Center. United States Department of Energy, Washington D.C., Tech. Rep., 2016. http://www.afdc.energy.gov/
66. NYSERDA, Large-Scale Renewable Energy Development in New York: Options and Assessment. New York State Energy Research and Development Authority, Tech. Rep., 2015
67. New York State of Opportunity, Reforming the Energy Vision (REV) (2018). https://static1.squarespace.com/static/576aad8437c5810820465107/t/5aec725baa4a99171e5890d4/1525445212467/REV-fm-fs-1-v8.pdf
68. State of California Energy Commission, California's 2030 Climate Commitment: Renewable Resources for Half of the State's Electricity by 2030. State of California Energy Commission, Tech. Rep., 2017
69. FERC, Electric Storage Participation in Markets Operated by Regional Transmission Organizations and Independent System Operators, Technical Report Docket Nos. RM16-23-000, Federal Energy Regulatory Commission (2016). https://www.ferc.gov/media/news-releases/2016/2016-4/11-17-16-E-1.asp#.XEJEBM9Kh0s
70. J.S. González, R. Lacal-Arántegui, A review of regulatory framework for wind energy in european union countries: current state and expected developments. Renew. Sustain. Energy Rev. **56**, 588–602 (2016)
71. M. Pacesila, S.G. Burcea, S.E. Colesca, Analysis of renewable energies in european union. Renew. Sustain. Energy Rev. **56**, 156–170 (2016)
72. EEA, Renewable energy in Europe – 2017 Update: Recent Growth and Knock-on Effects. European Environment Agency, Tech. Rep. No 23/2017 (2017)
73. B. Nykvist, M. Nilsson, Rapidly falling costs of battery packs for electric vehicles. Nat. Clim. Change **5**(4), 329–332 (2015)
74. D. Morris, China aims to push gas-powered cars out of the market (2017). http://fortune.com/2017/09/10/electric-cars-china/
75. B. Vlasic, N.E. Boudette, (2017) G.M. and Ford lay out plans to expand electric models. https://www.nytimes.com/2017/10/02/business/general-motors-electric-cars.html
76. Electric Vehicle Incentives (2018). https://www.tesla.com/support/incentives
77. J. Weiss, R. Hledik, M. Hagerty, W. Gorman, Electrification: emerging opportunities for utility growth. The Brattle Group, Tech. Rep., 2017
78. P.H. Andersen, J.A. Mathews, M. Rask, Integrating private transport into renewable energy policy: the strategy of creating intelligent recharging grids for electric vehicles. Energy Policy **37**(7), 2481–2486 (2009)
79. E. Sortomme, M.A. El-Sharkawi, Optimal scheduling of vehicle-to-grid energy and ancillary services. IEEE Trans. Smart Grid **3**(1), 351–359 (2012)

80. K. Clement-Nyns, E. Haesen, J. Driesen, The impact of charging plug-in hybrid electric vehicles on a residential distribution grid. IEEE Trans. Power Syst. **25**(1), 371–380 (2010)
81. E. Sortomme, M.M. Hindi, S.D.J. MacPherson, S.S. Venkata, Coordinated charging of plug-in hybrid electric vehicles to minimize distribution system losses. IEEE Trans. Smart Grid **2**(1), 198–205 (2011)
82. B.K. Sovacool, R.F. Hirsh, Beyond batteries: an examination of the benefits and barriers to plug-in hybrid electric vehicles (PHEVs) and a vehicle-to-grid (V2G) transition. Energy Policy **37**(3), 1095–1103 (2009)
83. M.D. Galus, M. Zima, G. Andersson, On integration of plug-in hybrid electric vehicles into existing power system structures. Energy Policy **38**(11), 6736–6745 (2010)
84. M.D. Galus, S. Koch, G. Andersson, Provision of load frequency control by PHEVs, controllable loads, and a cogeneration unit. IEEE Trans. Ind. Electron. **58**(10), 4568–4582 (2011)
85. F. Mwasilu, J.J. Justo, E.-K. Kim, T.D. Do, J.-W. Jung, Electricvehicles and smart grid interaction: a review on vehicle to grid and renewable energy sources integration. Renew. Sustain. Energy Rev. **34**, 501–516 (2014)
86. H. Lund, W. Kempton, Integration of renewable energy into the transport and electricity sectors through V2G. Energy Policy **36**(9), 3578–3587 (2008)
87. S.H. Low, A duality model of TCP and queue management algorithms. IEEE/ACM Trans. Networking **11**(4), 525–536 (2003)
88. D. Charypar, K. Nagel, An event-driven queue-based traffic flow microsimulation. Transp. Res. Rec. **2003**(1), 35–40 (2007)
89. A.M. Farid, A hybrid dynamic system model for multi-modal transportation electrification. IEEE Trans. Control Syst. Technol. **PP**(99), 1–12 (2016). http://dx.doi.org/10.1109/TCST.2016.2579602
90. A.M. Farid, Symmetrica: test case for transportation electrification research. Infrastruct. Complex. **2**(9), 1–10 (2015). http://dx.doi.org/10.1186/s40551-015-0012-9
91. T.J. van der Wardt, A.M. Farid, A hybrid dynamic system assessment methodology for multi-modal transportation-electrification. Energies **10**(5), 653 (2017). http://dx.doi.org/10.3390/en10050653
92. W.W. Hogan, Multiple market-clearing prices, electricity market design and price manipulation. Electr. J. **25**(4), 18–32 (2012). http://www.sciencedirect.com/science/article/pii/S1040619012000917
93. A. Garcia, L. Mili, J. Momoh, Modeling electricity markets: a brief introduction, in *Economic Market Design and Planning for Electric Power Systems* (Wiley, London, 2009), pp. 21–44. http://dx.doi.org/10.1002/9780470529164.ch2
94. D. Kirschen, G. Strbac, *Fundamentals of Power System Economics* (Wiley, West Sussex, 2004)
95. M. Shahidehpour, H. Yamin, Z. Li, *Frontmatter and Index* (Wiley, London, 2002)
96. A.S. Mishra, G. Agnihotri, N. Patidar, Transmission and wheeling service pricing: trends in deregulated electricity market. J. Adv. Eng. Sci. Sect. A **3**(1), 1–16 (2010)
97. Anonymous, Transforming Industries: Energy and Utilities. Ericsson, Tech. Rep., 2014. https://www.ericsson.com/assets/local/news/2014/10/gtwp-op-transforming-industries-aw-print.pdf
98. N. Rotering, M. Ilic, Optimal charge control of plug-in hybrid electric vehicles in deregulated electricity markets. IEEE Trans. Power Syst. **26**(3), 1021–1029 (2011)
99. X.P. Zhang, A framework for operation and control of smart grids with distributed generation, in *2008 IEEE Power and Energy Society General Meeting – Conversion and Delivery of Electrical Energy in the 21st Century* (IEEE, Pittsburgh, 2008), pp. 1–5. http://dx.doi.org/10.1109/PES.2008.4596345
100. Anonymous, Transactive energy (2017). http://www.gridwiseac.org/about/transactive_energy.aspx
101. F.F. Wu, K. Moslehi, A. Bose, Power system control centers: past, present, and future. Proc. IEEE **93**(11), 1890–1908 (2005)

102. J. Frazer, Smart Cities: Intelligent Transportation and Smart Grid Standards for Electrical and Lighting. Gridaptive, Tech. Rep., 2012
103. L. Gauchia, *Enabling Smart Grids: Energy Storage Technologies, Opportunities and Challenges* (IEEE Smart Grid, Piscataway, 2014)
104. A.M. Annaswamy, M. Amin, C.L. Demarco, T. Samad, J. Aho, G. Arnold, A. Buckspan, A. Cadena, D. Callaway, E. Camacho, M. Caramanis, A. Chakrabortty, A. Chakraborty, J. Chow, M. Dahleh, A.D. Dominguez-Garcia, D. Dotta, A.M. Farid, P. Flikkema, D. Gayme, S. Genc, M.G.I. Fisa, I. Hiskens, P. Houpt, G. Hug, P. Khargonekar, H. Khurana, A. Kiani, S. Low, J. McDonald, E. Mojica-Nava, A.L. Motto, L. Pao, A.Parisio, A. Pinder, M. Polis, M. Roozbehani, Z. Qu, N. Quijano, J. Stoustrup, in *IEEE Vision for Smart Grid Controls: 2030 and Beyond*, ed. by A.M. Annaswamy, M. Amin, C.L. Demarco, T. Samad (IEEE Standards Association, New York, 2013). http://www.techstreet.com/ieee/products/1859784
105. EPRI, Estimating the Costs and Benefits of the Smart Grid. EPRI, Palo Alto, Tech. Rep., 2011
106. V.C. Güngör, D. Sahin, T. Kocak, S. Ergüt, C. Buccella, S. Member, C. Cecati, G.P. Hancke, S. Member, Smart grid technologies: communication technologies and standards. IEEE Trans. Ind. Inf. **7**(4), 529–539 (2011)
107. V. Gungor, D. Sahin, T. Kocak, S. Ergut, C. Buccella, C. Cecati, G. Hancke, A survey on smart grid potential applications and communication requirements. IEEE Trans. Ind. Inf. **9**(1), 28–42 (2013)
108. B.-M. Hodge, A. Florita, K. Orwig, D. Lew, M. Milligan, A comparison of wind power and load forecasting error distributions, in *2012 World Renewable Energy Forum* (National Renewable Energy Laboratory, Denver, 2012)
109. A. Moreno-Munoz, J. de la Rosa, R. Posadillo, V. Pallares, Short term forecasting of solar radiation, in *2008 IEEE International Symposium on Industrial Electronics* (IEEE, Piscataway, 2008), pp. 1537–1541
110. A. von Meier, *Electric Power Systems: A Conceptual Introduction* (IEEE Press, Wiley-Interscience, Hoboken, 2006). http://ieeexplore.ieee.org/xpl/bkabstractplus.jsp?bkn=5238205
111. I. Atzeni, L.G. Ordóñez, G. Scutari, D.P. Palomar, J.R. Fonollosa, Demand-side management via distributed energy generation and storage optimization. IEEE Trans. Smart Grid **4**(2), 866–876 (2013)
112. P.P. Barker, R.W. De Mello, Determining the impact of distributed generation on power systems. I. radial distribution systems, in *2000 IEEE Power Engineering Society Summer Meeting*, vol. 3 (IEEE, Piscataway, 2000), pp. 1645–1656
113. P.T.R. Wang, C.R. Wanke, F.P. Wieland, Modeling time and space metering of flights in the National Airspace System, in *Proceedings of the 2004 Winter Simulation Conference*, vol. 2 (IEEE, Piscataway, 2004)
114. G. Pepermans, J. Driesen, D. Haeseldonckx, R. Belmans, W. D'haeseleer, Distributed generation: definition, benefits and issues. *Energy Policy* **33**(6), 787–798 (2005)
115. K. Ashton et al, That 'internet of things' thing. RFID J. **22**(7), 97–114 (2009)
116. S. Sarma, D. Brock, K. Ashton, The networked physical world – proposals for engineering the next generation of computing, commerce and automatic identification. Auto-ID Center MIT, Tech. Rep., 2000
117. P. Suresh, J.V. Daniel, V. Parthasarathy, R. Aswathy, A state of the art review on the internet of things (IoT) history, technology and fields of deployment, in *2014 International Conference on Science Engineering and Management Research (ICSEMR)* (IEEE, Piscataway, 2014), pp. 1–8
118. D. McFarlane, C. Carr, M. Harrison, A. McDonald, Auto-ID's Three R's: Rules and Recipes for Product Requirements, University of Cambridge – Institute for Manufacturing AUTO-ID Centre, Tech. Rep. CAM-AUTOID-WH-008 (2002)
119. A.I. Dashchenko, *Reconfigurable Manufacturing Systems and Transformable Factories* (Springer, Berlin, 2006)
120. R.M. Setchi, N. Lagos, Reconfigurability and reconfigurable manufacturing systems – state of the art review, in *Innovative Production Machines and Systems: Production Automation*

and Control State of the Art Review (Innovative Production Machines and Systems: Nework of Excellence, Cardiff, 2005), pp. 131–143
121. P. Leitao, V. Marik, P. Vrba, Past, present, and future of industrial agent applications. IEEE Trans. Ind. Inf. **9**(4), 2360–2372 (2013)
122. G.G. Meyer, K. Främling, J. Holmström, Intelligent products: a survey. Comput. Ind. **60**(3), 137–148 (2009)
123. D. McFarlane, V. Giannikas, A.C. Wong, M. Harrison, Product intelligence in industrial control: theory and practice. Annu. Rev. Control **37**(1), 69–88 (2013)
124. H.V. Brussel, J. Wyns, P. Valckenaers, L. Bongaerts, P. Peeters, Reference architecture for holonic manufacturing systems: PROSA. Comput. Ind. **37**, 255–274 (1998)
125. J. Wyns, Reference architecture for holonic manufacturing systems. Ph.D. Thesis, Catholic University of Leuven, 1999
126. H. Van Brussel, L. Bongaerts, J. Wyns, P. Valckenaers, T. Vanginderachter, A conceptual framework for holonic manufacturing: identification of manufacturing holons. J. Manuf. Syst. **18**(1), 35–52 (1999)
127. J.-L.L. Chirn, Developing a reconfigurable manufacturing control system – a holonic component based approach. Ph.D. Thesis, Univeristy of Cambridge, 2001
128. P. Leitão, F. Restivo, ADACOR: a holonic architecture for agile and adaptive manufacturing control. Comput. Ind. **57**(2), 121–130 (2006)
129. P. Leitao, An agile and adaptive holonic architecture for manufacturing control. Ph.D. Thesis, University of Porto, 2004
130. A.M. Farid, L. Ribeiro, An axiomatic design of a multi-agent reconfigurable mechatronic system architecture. IEEE Trans. Ind. Inf. **11**(5), 1142–1155 (2015). http://dx.doi.org/10.1109/TII.2015.2470528
131. A.M. Farid, L. Ribeiro, An axiomatic design of a multi-agent reconfigurable manufacturing system architecture, in *International Conference on Axiomatic Design*, Lisbon, 2014, pp. 1–8. http://engineering.dartmouth.edu/liines/resources/Conferences/IEM-C41.pdf
132. C. Donitzky, O. Roos, S. Sauty, A digital energy network: the internet of things and the smart grid (2014). https://www.intel.com/content/dam/www/public/us/en/documents/white-papers/iot-smart-grid-paper.pdf
133. Y. Saleem, N. Crespi, M.H. Rehmani, R. Copeland, Internet of things-aided smart grid: technologies, architectures, applications, prototypes, and future research directions. Preprint, arXiv:1704.08977 (2017)
134. Anonymous, Electric Power Monthly. U.S. Energy Information Administration, Tech. Rep., 2017
135. EPRI, *The Green Grid Energy Savings and Carbon Emissions Reductions Enabled by a Smart Grid* (Electric Power Research Institute, Palo Alto, 2008)
136. T. Ackermann, V. Knyazkin, Interaction between distributed generation and the distribution network: operation aspects, in *Transmission and Distribution Conference and Exhibition (Cat. No.02CH37377); Proceedings Asia Pacific Conference and Exhibition of the IEEE-Power Engineering Society on Transmission and Distribution*, vol. 2 (IEEE, Piscataway, 2002), pp. 1357–1362
137. W. Luan, D. Sharp, S. Lancashire, Smart grid communication network capacity planning for power utilities, in *Transmission and Distribution Conference and Exposition, 2010 IEEE PES* (IEEE, Piscataway, 2010), pp. 1–4
138. Y. Yang, F. Lambert, D. Divan, A survey on technologies for implementing sensor networks for power delivery systems, in *2007 IEEE Power Engineering Society General Meeting* (IEEE, Piscataway, 2007), pp. 1–8
139. L. Mainetti, L. Patrono, A. Vilei, Evolution of wireless sensor networks towards the internet of things: a survey, in *2011 19th International Conference onSoftware, Telecommunications and Computer Networks (SoftCOM)* (IEEE, Piscataway, 2011), pp. 1–6
140. A. Zanella, N. Bui, A. Castellani, L. Vangelista, M. Zorzi, Internet of things for smart cities. IEEE Internet Things J. **1**(1), 22–32 (2014)
141. H. Farhangi, The path of the smart grid. IEEE Power Energ. Mag. **8**(1), 18–28 (2010)

142. R. Singh, R. Uluski, J.T. Reilly, S. Martino, X. Lu, J. Wang, Foundational Report Series: Advanced Distribution Management Systems for Grid Modernization, DMS Industry Survey. Argonne National Laboratory (ANL), Tech. Rep., 2017
143. A. Daneels, W. Salter, What is SCADA? in *In International Conference on Accelerator and Large Experimental Physics Control Systems* Trieste (1999)
144. EPRI, *The Integrated Grid: Realizing the Full Value of Central and Distributed Energy Resources* (EPRI, Palo Alto, 2014)
145. R. Zafar, A. Mahmood, S. Razzaq, W. Ali, U. Naeem, K. Shehzad, Prosumer based energy management and sharing in smart grid. Renew. Sustain. Energy Rev. **82**, 1675–1684 (2018)
146. P. De Martini, K.M. Chandy, N. Fromer, Grid 2020: Towards a policy of renewable and distributed energy resources. Resnick Sustainability Institute Caltech, Tech. Rep., 2012
147. G. Joos, B.T. Ooi, D. McGillis, F.D. Galiana, R. Marceau, The potential of distributed generation to provide ancillary services, in *Proceedings of the 2000 Power Engineering Society Summer Meeting; Proceedings of the Conference on IEEE Power Engineering Society Transmission and Distribution*, vol. 3 (IEEE, Seattle, 2000), pp. 1762–1767
148. F.F. Wu, P.P. Varaiya, R.S. Hui, Smart grids with intelligent periphery: an architecture for the energy internet. Engineering **1**(4), 436–446 (2015)
149. J. Wen, Q. Wu, D. Turner, S. Cheng, J. Fitch, Optimal coordinated voltage control for power system voltage stability. IEEE Trans. Power Syst. **19**(2), 1115–1122 (2004)
150. G. Wu, S. Talwar, K. Johnsson, N. Himayat, K.D. Johnson, M2M: from mobile to embedded internet. IEEE Commun. Mag. **49**(4) (2011). http://dx.doi.org/10.1109/MCOM.2011.5741144
151. D. Bakken, A. Bose, K.M. Chandy, P.P. Khargonekar, A. Kuh, S. Low, A. von Meier, K. Poolla, P. Varaiya, F. Wu, GRIP–grids with intelligent periphery: control architectures for Grid2050$^{\pi}$, in *2011 IEEE International Conference on Smart Grid Communications (SmartGridComm)* (IEEE, Piscataway, 2011), pp. 7–12
152. N. Sharma, P. Sharma, D. Irwin, P. Shenoy, Predicting solar generation from weather forecasts using machine learning, in *2011 IEEE International Conference on Smart Grid Communications (SmartGridComm)* (IEEE, Piscataway, 2011), pp. 528–533.
153. M. Silva, H. Morais, Z. Vale, An integrated approach for distributed energy resource short-term scheduling in smart grids considering realistic power system simulation. Energy Convers. Manag. **64**, 273–288 (2012)
154. V.C. Gungor, B. Lu, G.P. Hancke, Opportunities and challenges of wireless sensor networks in smart grid. IEEE Trans. Ind. Electron. **57**(10), 3557–3564 (2010)
155. M. Uyterlinde, E. Van Sambeek, E. Cross, W. Jörß, B.P. Löffler, P. Morthorst, B.H. Jørgensen, *Decentralised Generation: Development of EU Policy* (Energy Research Centre of the Netherlands (ECN), Petten, 2002)
156. S. Sarma, Towards the 5 cent Tag, in *White Paper* (2001), pp. 1–19.
157. X. Fang, S. Misra, G. Xue, D. Yang, Smart grid—the new and improved power grid: a survey. IEEE Commun. Surv. Tutorials **14**(4), 944–980 (2012)
158. G.S. Vassell, The northeast blackout of 1965. Public Util. Fortnightly (US) **126**(8) (1990)
159. Wikipedia Contributors, Northeast blackout of 1965—Wikipedia, The Free Encyclopedia, 2018. https://en.wikipedia.org/w/index.php?title=Northeast_blackout_of_1965&oldid=839900966
160. J. Lin, B. Zhu, P. Zeng, W. Liang, H. Yu, Y. Xiao, Monitoring power transmission lines using a wireless sensor network. Wirel. Commun. Mob. Comput. **15**(14), 1799–1821 (2015)
161. Q. Ou, Y. Zhen, X. Li, Y. Zhang, L. Zeng, Application of internet of things in smart grid power transmission, in *2012 Third FTRA International Conference on Mobile, Ubiquitous, and Intelligent Computing* (IEEE, Piscataway, 2012), pp. 96–100
162. Y. Ebata, H. Hayashi, Y. Hasegawa, S. Komatsu, K. Suzuki, Development of the intranet-based scada (supervisory control and data acquisition system) for power system, in *2000 IEEE Power Engineering Society Winter Meeting. Conference Proceedings*, vol. 3 (IEEE, Piscataway, 2000), pp. 1656–1661

163. N. Zhou, D. Meng, Z. Huang, G. Welch, Dynamic state estimation of a synchronous machine using pmu data: a comparative study. IEEE Trans. Smart Grid **6**(1), 450–460 (2015)
164. W. Li and J. Korczynski, A reliability based approach to transmission maintenance planning and its application in BC hydro system, in *2001 Power Engineering Society Summer Meeting. Conference Proceedings*, vol. 1 (IEEE, Piscataway, 2001), pp. 510–515
165. C. Opfimizdion, Static state estimation in electric power systems. Proc. IEEE **62**(7), 972–982 (1973)
166. A. Monticelli, Electric power system state estimation. Proc. IEEE **88**(2), 262–282 (2000)
167. M. Zima-Bockarjova, M. Zima, G. Andersson, Analysis of the state estimation performance in transient conditions. IEEE Trans. Power Syst. **26**(4), 1866–1874 (2011)
168. N.R. Shivakumar, A. Jain, A review of power system dynamic state estimation techniques, in *2008 Joint International Conference on Power System Technology and IEEE Power India Conference* (IEEE, New Delhi, 2008), pp. 1–6. http://dx.doi.org/10.1109/ICPST.2008.4745312
169. M.D. Ilic, L. Xie, U.A. Khan, Modeling future cyber-physical energy systems, in *2008 IEEE Power and Energy Society General Meeting - Conversion and Delivery of Electrical Energy in the 21st Century* (IEEE, Piscataway, 2008), pp. 1–9
170. J. Bertsch, M. Zima, A. Suranyi, C. Carnal, C. Rehtanz, Experiences with and perspectives of the system for wide area monitoring of power systems, in *CIGRE/IEEE PES International Symposium Quality and Security of Electric Power Delivery Systems, 2003. CIGRE/PES 2003* (IEEE, Piscataway, 2003), pp. 5–9
171. V. Terzija, G. Valverde, D. Cai, P. Regulski, V. Madani, J. Fitch, S. Skok, M.M. Begovic, A. Phadke, Wide-area monitoring, protection, and control of future electric power networks. Proc. IEEE **99**(1), 80–93 (2011)
172. A. Bose, Smart transmission grid applications and their supporting infrastructure. IEEE Trans. Smart Grid **1**(1), 11–19 (2010)
173. D.-K.N.A. Editors, Minimizing IoT Sensor Node Power Consumption. Web Article, 2017
174. M.S. Amin, B.F. Wollenberg, Toward a smart grid: power delivery for the 21st century. IEEE Power Energ. Mag. **3**(5), 34–41 (2005). http://dx.doi.org/10.1109/MPAE.2005.1507024
175. A.G. Phadke, Synchronized phasor measurements in power systems. IEEE Comput. Appl. Power **6**(2), 10–15 (1993)
176. R. Burnett, M. Butts, P. Sterlina, Power system applications for phasor measurement units. IEEE Comput. Appl. Power **7**(1), 8–13 (1994)
177. A.G. Phadke, J.S. Thorp, *Synchronized Phasor Measurements and Their Applications* (Springer, New York, 2008)
178. D.A. Haughton, G.T. Heydt, Synchronous measurements in power distribution systems, in *Control and Optimization Methods for Electric Smart Grids*, ed. by M.D. Ilic, A. Chakrabortty (Springer, New York, 2012), pp. 295–312
179. S. Horowitz, A. Phadke, B. Renz, The future of power transmission. IEEE Power Energy Mag. **8**(2), 34–40 (2010)
180. M.D. Ilic, L. Xie, U.A. Khan, J.M.F. Moura, Modeling of future cyber and physical energy systems for distributed sensing and control. IEEE Trans. Syst. Man Cybern. Part A Syst. Hum. **40**(4), 825–838 (2010)
181. A. Gomez-Exposito, A. Abur, A. De La Villa Jaen, C. Gomez-Quiles, A multilevel state estimation paradigm for smart grids. Proc. IEEE **99**(6), 952–976 (2011). http://dx.doi.org/10.1109/JPROC.2011.2107490
182. K. Zhu, M. Chenine, L. Nordstrom, ICT architecture impact on wide area monitoring and control systems' reliability. IEEE Trans. Power Delivery **26**(4), 2801–2808 (2011)
183. A. Gomez-Exposito, A. Abur, A. de la Villa Jaen, C. Gomez-Quiles, A multilevel state estimation paradigm for smart grids. Proc. IEEE **99**(6), 952–976 (2011)
184. A.-A. Fish, S. Chowdhury, S.P. Chowdhury, Optimal PMU placement in a power network for full system observability, in *2011 IEEE Power and Energy Society General Meeting* (IEEE, Piscataway, 2011), pp. 1–8
185. A. Kumar, S. Chakrabarti, Ann-based hybrid state estimation and enhanced visualization of power systems, in *ISGT2011-India* (IEEE, Piscataway, 2011), pp. 78–83

186. D. Tholomier, H. Kang, B. Cvorovic, Phasor measurement units: functionality and applications, in *2009 Power Systems Conference* (IEEE Computer Society, Clemson, 2009), p. 1. http://dx.doi.org/10.1109/PSAMP.2009.5262468
187. S. Chakrabarti, E. Kyriakides, G. Ledwich, A. Ghosh, Inclusion of PMU current phasor measurements in a power system state estimator. IET Gener. Transm. Distrib. **4**(10), 1104–1115 (2010)
188. B. Gou, Generalized integer linear programming formulation for optimal PMU placement. IEEE Trans. Power Syst. **23**(3), 1099–1104 (2008)
189. A. Glazunova, I. Kolosok, E. Korkina, PMU placement on the basis of Scada measurements for fast load flow calculation in electric power systems, in *2009 IEEE Bucharest PowerTech* (IEEE, Piscataway, 2009), pp. 1–6
190. N.M. Manousakis, G.N. Korres, Observability analysis for power systems including conventional and phasor measurements, in *7th Mediterranean Conference and Exhibition on Power Generation, Transmission, Distribution and Energy Conversion* (IET, Agia Napa, 2010), pp. 1–8
191. M. Begovic, D. Novosel, B. Djokic, Issues related to the implementation of synchrophasor measurements, in *Proceedings of the 41st Annual Hawaii International Conference on System Sciences* (IEEE, Piscataway, 2008), p. 164
192. M. Berg, J. Stamp, *A Reference Model for Control and Automation Systems in Electric Power* (Sandia National Laboratories, Albuquerque, 2005)
193. L.L. Grigsby. Power System Stability and Control (CRC press, 2016)
194. B. Fardanesh, Future trends in power system control. IEEE Comput. Appl. Power **15**(3), 24–31 (2002)
195. N.G. Hingorani, L. Gyugyi, *Understanding FACTS: Concepts and Technology of Flexible AC Transmission Systems* (IEEE Press, New York, 2000). http://ieeexplore.ieee.org/xpl/bkabstractplus.jsp?bkn=5264253
196. H. Glavitsch, J. Stoffel, Automatic generation control. Int. J. Electr. Power Energy Syst. **2**(1), 21–28 (1980)
197. N. Jaleeli, L.S. VanSlyck, D.N. Ewart, L.H. Fink, A.G. Hoffmann, Understanding automatic generation control. IEEE Trans. Power Syst. **7**(3), 1106–1122 (1992)
198. A.N. Venkat, I.A. Hiskens, J.B. Rawlings, S.J. Wright, Distributed mpc strategies with application to power system automatic generation control. IEEE Trans. Control Syst. Technol. **16**(6), 1192–1206 (2008)
199. FERC, Third-party provision of ancillary services; accounting and financial reporting for new electric storage technologies (2013). https://www.ferc.gov/whats-new/comm-meet/2013/071813/E-22.pdf
200. T. Li, Z. Xiao, M. Huang, J. Yu, J. Hu, Control system simulation of microgrid based on ip and multi-agent, *2010 International Conference on Information, Networking and Automation (ICINA)* (IEEE, Piscataway, 2010), pp. V1-235–V1-239
201. M.E. Elkhatib, R.E. Shatshat, M.M.A. Salama, Decentralized reactive power control for advanced distribution automation systems. IEEE Trans. Smart Grid **3**(3), 1482–1490 (2012)
202. A.K. Mohanty, A.K. Barik, Power system stability improvement using facts devices. Int. J. Mod. Eng. Res. **1**(2), 666–672 (2011)
203. E.A. Rogers, *The Energy Savings Potential of Smart Manufacturing* (American Council for an Energy-Efficient Economy, Washington, 2014)
204. G.A. Munoz-Hernandez, S.P. Mansoor, D.I. Jones, *Modelling and Controlling Hydropower Plants* (Springer, London, 2013)
205. W.N. Lubega, An engineering systems approach to the modeling and analysis of the energy-water nexus. Master's Thesis, Masdar Institute of Science and Technology, 2014
206. W.N. Lubega, A.M. Farid, Quantitative engineering systems model and analysis of the energy-water nexus. Appl. Energy **135**(1), 142–157 (2014). http://dx.doi.org/10.1016/j.apenergy.2014.07.101
207. W.N. Lubega, A.M. Farid, A reference system architecture for the energy-water nexus. IEEE Sys. J. **PP**(99), 1–11 (2014). http://dx.doi.org/10.1109/JSYST.2014.2302031

208. W.N. Lubega, A. Santhosh, A.M. Farid, K. Youcef-Toumi, Opportunities for Integrated Energy and Water Management in the GCC – A Keynote Paper, in *EU-GCC Renewable Energy Policy Experts' Workshop* (Masdar Institute, Abu Dhabi, 2013), pp. 1–33. http://www.grc.net/data/contents/uploads/Opportunities_for_Integrated_Energy_and_Water_Management_in_the_GCC_5874.pdf
209. W.N. Lubega, A. Santhosh, A.M. Farid, K. Youcef-Toumi, An integrated energy and water market for the supply side of the energy-water nexus in the engineered infrastructure, in *ASME 2014 Power Conference*, Baltimore (2014), pp. 1–11. http://dx.doi.org/10.1115/POWER2014-32075
210. W.N. Lubega, A.M. Farid, An engineering systems sensitivity analysis model for holistic energy-water nexus planning, in *ASME 2014 Power Conference*, Baltimore (2014), pp. 1–10. http://dx.doi.org/10.1115/POWER2014-32076
211. A.M. Farid, W.N. Lubega, Powering and watering agriculture: application of energy-water nexus planning, in *GHTC 2013: IEEE Global Humanitarian Technology Conference*, Silicon Valley (2013), pp. 1–6. http://dx.doi.org.libproxy.mit.edu/10.1109/GHTC.2013.6713689
212. A. Santhosh, A.M. Farid, A. Adegbege, K. Youcef-Toumi, Simultaneous co-optimization for the economic dispatch of power and water networks, in *The 9th IET International Conference on Advances in Power System Control, Operation and Management*, Hong Kong (2012), pp. 1–6. http://dx.doi.org/10.1049/cp.2012.2148
213. A. Santhosh, A.M. Farid, K.Youcef-Toumi, The impact of storage facilities on the simultaneous economic dispatch of power and water networks limited by ramping rates, in *IEEE International Conference on Industrial Technology*, Cape Town (2013), pp. 1–6. http://dx.doi.org/10.1109/ICIT.2013.6505794
214. A. Santhosh, A.M. Farid, K. Youcef-Toumi, The impact of storage facility capacity and ramping capabilities on the supply side of the energy-water nexus. Energy **66**(1), 1–10, 2014. http://dx.doi.org/10.1016/j.energy.2014.01.031
215. W. Hickman, A. Muzhikyan, A.M. Farid, The synergistic role of renewable energy integration into the unit commitment of the energy water nexus. Renew. Energy **108**, 220–229 (2017). https://dx.doi.org/10.1016/j.renene.2017.02.063
216. A.M. Farid, W.N. Lubega, W. Hickman, Opportunities for energy-water nexus management in the Middle East and North Africa. Elementa **4**(134), 1–17 (2016). http://dx.doi.org/10.12952/journal.elementa.000134
217. B.C. Klein, A. Bonomi, R.M. Filho, Integration of microalgae production with industrial biofuel facilities: a critical review. Renew. Sustain. Energy Rev. **82**, 1376–1392 (2018). http://www.sciencedirect.com/science/article/pii/S1364032117305701
218. D. Murrant, A. Quinn, L. Chapman, The water-energy nexus: future water resource availability and its implications on UK thermal power generation. Water Environ. J. **29**(3), 307–319 (2015)
219. M. Wakeel, B. Chen, Energy consumption in urban water cycle. Energy Procedia **104**, 123–128 (2016). http://www.sciencedirect.com/science/article/pii/S1876610216315776
220. M. Bekchanov, A. Sood, A. Pinto, M. Jeuland, Systematic review of water-economy modeling applications. J. Water Resour. Plan. Manag. **143**(8), 04017037 (2017)
221. J. Macknick, R. Newmark, G. Heath, K.C. Hallett, Operational water consumption and withdrawal factors for electricity generating technologies: a review of existing literature. Environ. Res. Lett. **7**(4), 045802 (2012)
222. J. Macknick, S. Sattler, K. Averyt, S. Clemmer, J. Rogers, The water implications of generating electricity: water use across the united states based on different electricity pathways through 2050. Environ. Res. Lett. **7**(4), 045803 (2012)
223. J. Rogers, K. Averyt, S. Clemmer, M. Davis, F. Flores-Lopez, D. Kenney, J. Macknick, N. Madden, J. Meldrum, S. Sattler, E. Spanger-Siegfried, Water-Smart Power: Strengthening the U.S. Electricity System in a Warming World. Union for Concerned Scientists, Cambridge, Tech. Rep., 2013
224. K. Averyt, J. Fisher, A. Huber-Lee, A. Lewis, J. Macknick, N. Madden, J. Rogers, S. Tellinghuisen, Freshwater Use by US Power Plants: Electricity's Thirst for a Precious Resource. Union of Concerned Scientists, Cambridge, Tech. Rep., 2011

225. V.C. Tidwell, J.Macknick, K. Zemlick, J. Sanchez, T. Woldeyesus, Transitioning to zero freshwater withdrawal in the U.S. for thermoelectric generation. Appl. Energy **131**, 508–516 (2014). http://linkinghub.elsevier.com/retrieve/pii/S0306261913009215
226. B.D.H. Kiran, M.S. Kumari, Demand response and pumped hydro storage scheduling for balancing wind power uncertainties: a probabilistic unit commitment approach. Int. J. Electr. Power Energy Syst. **81**, 114–122 (2016)
227. W.A. Hussien, F.A. Memon, D.A. Savic, An integrated model to evaluate water-energy-food nexus at a household scale. Environ. Model. Softw. **93**, 366–380 (2017). http://www.sciencedirect.com/science/article/pii/S1364815216306594
228. D. ZhaoYang, W. Kit Po, M. Ke, L. Fengji, Y. Fang, Z. Junhua, Wind power impact on system operations and planning, in *2010 IEEE Power and Energy Society General Meeting* (IEEE, Piscataway, 2010), pp. 1–5
229. Z. Dong, K.P. Wong, K. Meng, F. Luo, F. Yao, J. Zhao, Wind power impact on system operations and planning, in *IEEE PES General Meeting* (IEEE, Piscataway, 2010), pp. 1–5
230. L. Xie, P.M.S. Carvalho, L.A.F.M. Ferreira, J. Liu, B.H. Krogh, N. Popli, M.D. Ilić, Wind integration in power systems: operational challenges and possible solutions. Proc. IEEE **99**(1), 214–232 (2011)
231. W.F. Pickard, A.Q. Shen, N.J. Hansing, Parking the power: strategies and physical limitations for bulk energy storage in supply-demand matching on a grid whose input power is provided by intermittent sources. Renew. Sustain. Energy Rev. **13**(8), 1934–1945 (2009). http://linkinghub.elsevier.com/retrieve/pii/S1364032109000562http://dx.doi.org/10.1016/j.rser.2009.03.002
232. L. Xie, M.D. Ilic, M.D. Ili, Model predictive economic/environmental dispatch of power systems with intermittent resources, in *2009 Power and Energy Society General Meeting* (IEEE, Piscataway, 2009), pp. 1–6
233. M. Chaabene, M. Annabi, Dynamic thermal model for predicting solar plant adequate energy management. Energy Convers. Manag. **39**(3–4), 349–355 (1998)
234. H. Zheng, A. Kusiak, Prediction of wind farm power ramp rates: a data-mining approach. J. Sol. Energy Eng. **131**(3), 031011(2009)
235. R. Baños, F. Manzano-Agugliaro, F. Montoya, C. Gil, A. Alcayde, J. Gómez, Optimization methods applied to renewable and sustainable energy: a review. Renew. Sustain. Energy Rev. **15**(4), 1753–1766 (2011)
236. J. Wu, S. Yang, Optimal movement-assisted sensor deployment and its extensions in wireless sensor networks. Simul. Model. Pract. Theory **15**(4), 383–399 (2007). http://dx.doi.org/10.1016/j.simpat.2006.11.006
237. G. Giebel, R. Brownsword, G. Kariniotakis, M. Denhard, C. Draxl, *The State-of-the-Art in Short-Term Prediction of Wind Power: A Literature Overview*, 2nd edn. (Anemos.Plus, Roskilde, 2011)
238. A.E. Curtright, J. Apt, The character of power output from utility-scale photovoltaic systems. Prog. Photovolt. Res. Appl. **16**(3), 241–247 (2008)
239. J. Apt, A. Curtright, The spectrum of power from utility-scale wind farms and solar photovoltaic arrays. Carnegie Mellon Electricity Industry Center Working Paper, Pittsburgh, Tech. Rep., 2007
240. A.M. Foley, P.G. Leahy, A. Marvuglia, E.J. McKeogh, Current methods and advances in forecasting of wind power generation. Renew. Energy **37**(1), 1–8 (2012)
241. C.W. Potter, A. Archambault, K. Westrick, Building a smarter smart grid through better renewable energy information, in *2009 IEEE/PES Power Systems Conference and Exposition* (IEEE, Piscataway, 2009), pp. 1–5
242. S.S. Soman, H. Zareipour, O. Malik, P. Mandal, A review of wind power and wind speed forecasting methods with different time horizons, in *North American Power Symposium (NAPS), 2010* (IEEE, Piscataway, 2010), pp. 1–8
243. B.-M. Hodge, M. Milligan, Wind power forecasting error distributions over multiple timescales, in *2011 IEEE Power and Energy Society General Meeting* (IEEE, Piscataway, 2011), pp. 1–8

244. V. Kostylev, A. Pavlovski et al., Solar power forecasting performance–towards industry standards, in *1st International Workshop on the Integration of Solar Power into Power Systems*, Aarhus (2011)
245. A. Muzhikyan, A.M. Farid, K. Youcef-Toumi, An enterprise control assessment method for variable energy resource induced power system imbalances. Part 1: methodology. IEEE Trans. Ind. Electron. **62**(4), 2448–2458 (2015). http://dx.doi.org/10.1109/TIE.2015.2395391
246. A. Muzhikyan, A.M. Farid, K. Youcef-Toumi, An enterprise control assessment method for variable energy resource induced power system imbalances. Part 2: results. IEEE Trans. Ind. Electron. **62**(4), 2459–2467 (2015). http://dx.doi.org/10.1109/TIE.2015.2395380
247. A. Muzhikyan, A.M. Farid, K. Youcef-Toumi, Relative merits of load following reserves and energy storage market integration towards power system imbalances. Int. J. Electr. Power Energy Syst. **74**(1), 222–229 (2016). http://dx.doi.org/10.1016/j.ijepes.2015.07.013
248. A. Muzhikyan, A.M. Farid, T. Mezher, The impact of wind power geographical smoothing on operating reserve requirements, in *IEEE American Control Conference* (IEEE, Boston, 2016), pp. 1–6. http://dx.doi.org/10.1109/ACC.2016.7526593
249. A. Muzhikyan, A.M. Farid, K. Youcef-Toumi, An a priori analytical method for determination of operating reserves requirements. Int. J. Energy Power Syst. **86**(3), 1–11 (2016). http://dx.doi.org/10.1016/j.ijepes.2016.09.005
250. B. Jiang, A. Muzhikyan, A.M. Farid, K. Youcef-Toumi, Demand side management in power grid enterprise control—a comparison of industrial and social welfare approaches. Appl. Energy **187**, 833–846 (2017). http://dx.doi.org/10.1016/j.apenergy.2016.10.096
251. G. van Welie, State of the grid: 2017, ISO on background (2017). https://www.iso-ne.com/static-assets/documents/2017/01/20170130_stateofgrid2017_presentation_pr.pdf
252. S. An, Q. Li, T. Gedra, Natural gas and electricity optimal power flow, in *2003 IEEE PES Transmission and Distribution Conference and Exposition*, vol. 1 (IEEE, Piscataway, 2003), pp. 138–143
253. M. Shahidehpour, Y. Fu, T. Wiedman, Impact of natural gas infrastructure on electric power systems. Proc. IEEE **93**(5), 1042–1056 (2005)
254. H. Chen, R. Baldick, Optimizing short-term natural gas supply portfolio for electric utility companies. IEEE Trans. Power Syst. **22**(1), 232–239 (2007)
255. J. Munoz, N. Jimenez-Redondo, J. Perez-Ruiz, J. Barquin, Natural gas network modeling for power systems reliability studies, in *2003 IEEE Bologna PowerTech Conference* (IEEE, Piscataway, 2003)
256. C. Unsihuay, J.W.M. Lima, A.C.Z.D. Souza, Modeling the integrated natural gas and electricity optimal power flow, in *IEEE Power Engineering Society General Meeting* (IEEE, Piscataway, 2007), pp. 1–7
257. T. Li, M. Eremia, M. Shahidehpour, M. Shahidepour, Interdependency of natural gas network and power system security. IEEE Trans. Power Syst. **23**(4), 1817–1824 (2008)
258. ICF, Assessment of New England's Natural Gas Pipeline Capacity to Satisfy Short and Near-term Power Generation Needs. ICF International for ISO-NE Planning Advisory Committee, Tech. Rep., 2012
259. S.N. Malik, O.R. Sukhera, Management of natural gas resources and search for alternative renewable energy resources: a case study of Pakistan. Renew. Sustain. Energy Rev. **16**(2), 1282–1290 (2012). http://dx.doi.org/10.1016/j.rser.2011.10.003
260. C. Davies, A. Goller, Ensuring future natural gas availability, in *White Paper Prepared for the 2013 MITei Symposium on Interdependency of Gas and Electricity Systems* (2013)
261. S. Adamson, R. Tabors, Pricing short-term gas availability in power markets, in *MITEI Symposium on Interdependency of Natural Gas and Electricity Systems* (2013)
262. C.M. Correa-Posada, P. Sanchez-Martin, Security-constrained optimal power and natural-gas flow. IEEE Trans. Power Syst. **29**(4), 1780–1787 (2013)
263. T. Oleson, *America's Increasing Reliance on Natural Gas: Benefits and Risks of a Methane Economy* (American Geosciences Institute, Alexandria, 2015)
264. D.F. Opila, A.M. Zeynu, I.A. Hiskens, Wind farm reactive support and voltage control, in *2010 IREP Symposium Bulk Power System Dynamics and Control - VIII (IREP)* (IEEE, Piscataway,2010), pp. 1–10

265. IEEE PES Wind Plant Collector System Design Working Group, Wind power generators for wind power plants (2018). http://www.site.uottawa.ca/~rhabash/ELG4126WindGenerators.pdf
266. M. Pelletier, M. Phethean, S. Nutt, Grid code requirements for artificial inertia control systems in the new zealand power system, in *2012 IEEE Power and Energy Society General Meeting* (IEEE, Piscataway, 2012), pp. 1–7
267. P. Fairley, Can Synthetic Inertia from Wind Power Stabilize Grids? IEEE Spectrum, Tech. Rep., 2016
268. E. Ortjohann, M. Lingemann, A. Mohd, W. Sinsukthavorn, A. Schmelter, N. Hamsic, D. Morton, A general architecture for modular smart inverters, in *2008 IEEE International Symposium on Industrial Electronics* (IEEE, Piscataway, 2008), pp. 1525–1530. http://dx.doi.org/10.1109/ISIE.2008.4676908
269. K. Zipp, What is a Smart Solar Inverter? Solar Power World Online, Tech. Rep., 2014. https://www.solarpowerworldonline.com/2014/01/smart-solar-inverter/
270. L. Morales-Velazquez, R. de Jesus Romero-Troncoso, G. Herrera-Ruiz, D. Morinigo-Sotelo, R.A. Osornio-Rios, Smart sensor network for power quality monitoring in electrical installations. Measurement **103**, 133–142 (2017)
271. M. Geberslassie, B. Bitzer, Future SCADA systems for decentralized distribution systems, in *45th International Universities Power Engineering Conference UPEC2010* (IEEE Computer Society, Cardiff, 2010), pp. 1–4
272. V.M. Igure, S.A. Laughter, R.D. Williams, Security issues in scada networks. Comput. Secur. **25**(7), 498–506 (2006)
273. M. Weiss, A. Helfenstein, F. Mattern, T. Staake, Leveraging smart meter data to recognize home appliances, in *2012 IEEE International Conference on Pervasive Computing and Communications (PerCom)* (IEEE, Piscataway, 2012), pp. 190–197
274. S.S.S.R. Depuru, L. Wang, V. Devabhaktuni, Smart meters for power grid: challenges, issues, advantages and status. Renew. Sustain. Energy Rev. **15**(6), 2736–2742 (2011)
275. R. Morello, S.C. Mukhopadhyay, Z. Liu, D. Slomovitz, S.R. Samantaray, Advances on sensing technologies for smart cities and power grids: a review. IEEE Sensors J. **17**(23), 7596–7610 (2017)
276. R.R. Mohassel, A. Fung, F. Mohammadi, K. Raahemifar, A survey on advanced metering infrastructure. Int. J. Electr. Power Energy Syst. **63**, 473–484 (2014)
277. R. Bayindir, I. Colak, G. Fulli, K. Demirtas, Smart grid technologies and applications. Renew. Sustain. Energy Rev. **66**, 499–516 (2016)
278. L. Li, H. Xiaoguang, H. Jian, H. Ketai, Design of new architecture of AMR system in smart grid, in *2011 6th IEEE Conference on Industrial Electronics and Applications (ICIEA)* (IEEE, Piscataway, 2011), pp. 2025–2029
279. Anonymous, Annual Electric Power Industry Report. U.S. Energy Information Administration, Tech. Rep., 2016
280. U.E.I. Administration, How many smart meters are installed in the United States, and who has them? (2017). https://www.eia.gov/tools/faqs/faq.php?id=108&t=3
281. X. Zhou, P. Xiang, Y. Ma, Z. Gao, Y. Wu, J. Yin, X. Xu, An overview on distribution automation system, in *2016 Chinese Control and Decision Conference (CCDC)* (IEEE, Piscataway, 2016), pp. 3667–3671
282. M.S. Thomas, S. Arora, V.K. Chandna, Distribution automation leading to a smarter grid, in *2011 IEEE PES Innovative Smart Grid Technologies-India (ISGT India)* (IEEE, Piscataway, 2011), pp. 211–216
283. R.J. de Groot, J. Morren, J.G. Slootweg, Smart integration of distribution automation applications, in *2012 3rd IEEE PES International Conference and Exhibition on Innovative Smart Grid Technologies (ISGT Europe)* (IEEE, Piscataway, 2012), pp. 1–7
284. C. Wilson, T. Hargreaves, R. Hauxwell-Baldwin, Smart homes and their users: a systematic analysis and key challenges. Pers. Ubiquit. Comput. **19**(2), 463–476 (2015)
285. R. Missaoui, H. Joumaa, S. Ploix, S. Bacha, Managing energy smart homes according to energy prices: analysis of a building energy management system. Energy Build. **71**, 155–167 (2014)

286. S. Mennicken, J. Vermeulen, E.M. Huang, From today's augmented houses to tomorrow's smart homes: new directions for home automation research, in *Proceedings of the 2014 ACM International Joint Conference on Pervasive and Ubiquitous Computing* (ACM, New York, 2014), pp. 105–115
287. P.H. Shaikh, N.B.M. Nor, P. Nallagownden, I. Elamvazuthi, T. Ibrahim, A review on optimized control systems for building energy and comfort management of smart sustainable buildings. Renew. Sustain. Energy Rev. **34**, 409–429 (2014)
288. D. Lazos, A.B. Sproul, M. Kay, Optimisation of energy management in commercial buildings with weather forecasting inputs: a review. Renew. Sustain. Energy Rev. **39**, 587–603 (2014)
289. M. Brettel, N. Friederichsen, M. Keller, M. Rosenberg, How virtualization, decentralization and network building change the manufacturing landscape: an industry 4.0 perspective. Int. J. Mech. Ind. Sci. Eng. **8**(1), 37–44 (2014)
290. F. Shrouf, J. Ordieres, G. Miragliotta, Smart factories in industry 4.0: a review of the concept and of energy management approached in production based on the internet of things paradigm, in *2014 IEEE International Conference on Industrial Engineering and Engineering Management (IEEM)* (IEEE, Piscataway, 2014), pp. 697–701
291. K. Bojanczyk, Home Energy Management Systems: Vendors, Technologies and Opportunities, 2013–2017. GTM Research, Tech. Rep., 2013
292. R.J. Meyers, E.D. Williams, H.S. Matthews, Scoping the potential of monitoring and control technologies to reduce energy use in homes. Energy Build. **42**(5), 563–569 (2010)
293. R. Matulka, Energy efficiency tricks to stop your energy bill from haunting you (2013). https://www.energy.gov/articles/energy-efficiency-tricks-stop-your-energy-bill-haunting-you
294. A.E.C. Inc., Take control and save: phantom loads (2012). https://www.takecontrolandsave.coop/documents/PhantomLoad.pdf
295. E. Huffstetler, Learn how much a phantom load is costing you (2018). https://www.thebalance.com/what-is-a-phantom-load-and-how-much-is-it-costing-you-1388190
296. J. Chu, 3 easy tips to reduce your standby power loads (2012). https://www.energy.gov/energysaver/articles/3-easy-tips-reduce-your-standby-power-loads
297. S. Little, Why home automation is essential for energy efficiency (2015). https://freshome.com/home-automation-essential-energy-efficiency/
298. C.O. Adika, L. Wang, Autonomous appliance scheduling for household energy management. IEEE Trans. Smart Grid **5**(2), 673–682 (2014)
299. M. Saltzman, Light up your house with LEDs, and other tips to slash your electric bill (2017). https://www.usatoday.com/story/tech/columnist/saltzman/2017/11/25/light-up-your-house-leds-and-other-energy-saving-tips/894537001/
300. Anonymous-DOE, How Energy-Efficient Light Bulbs Compare with Traditional Incandescents. U.S. Department of Energy, Tech. Rep., 2018. https://www.energy.gov/energysaver/save-electricity-and-fuel/lighting-choices-save-you-money/how-energy-efficient-light
301. Anonymous, Energy Star Most Efficient 2018—Medium, Large, and x-Large Refrigerators. Energy Star, Tech. Rep., 2018. https://www.energystar.gov/most-efficient/me-certified-refrigerators
302. J. Gehrt, Intelligent HVAC systems through smart building technology (2017). https://facilityexecutive.com/2017/01/staying-cool-with-intelligent-hvac-systems/
303. Appkettle (2018). https://www.myappkettle.com/
304. G. Caldecott, Appkettle—review. http://www.radiotimes.com/news/2017-09-14/appkettle-review/
305. R. Baldwin, The World Now has a Smart Toaster. Engadget, Tech. Rep., 2018. https://www.engadget.com/2017/01/04/griffin-connects-your-toast-to-your-phone/
306. B. Heater, Smart Toasters are Here. https://techcrunch.com/2017/01/07/toaster/
307. G. Technology, Griffin technology unveils griffin home, a collection of smart, apppowered appliances that simplify and enhance everyday routines at CES 2017. https://press.griffintechnology.com/release/griffin-technology-unveils-griffin-home-a-collection-of-smart-apppowered-appliances-that-simplify-and-enhance-everyday-routines-at-ces-2017/
308. Anonymous, June. June Life Inc., Tech. Rep., 2018. https://juneoven.com/

309. Anonymous, Dyson Pure Hot+Cool Link^{TM} Purifier. Dyson, Tech. Rep., 2018. https://www.dyson.com/purifiers/dyson-pure-hot-cool-link-overview.html
310. Anonymous, Dyson Link App. Dyson, Tech. Rep., 2018. https://www.dyson.com/purifiers/dyson-pure-hot-cool-link-dyson-link-app.html
311. K. Chua, S. Chou, W. Yang, J. Yan, Achieving better energy-efficient air conditioning—a review of technologies and strategies. Appl. Energy **104**, 87–104 (2013)
312. B. Riangvilaikul, S. Kumar, An experimental study of a novel dew point evaporative cooling system. Energy Build. **42**(5), 637–644 (2010)
313. M. Gupta, S.S. Intille, K. Larson, Adding gps-control to traditional thermostats: an exploration of potential energy savings and design challenges, in *International Conference on Pervasive Computing* (Springer, Berlin, 2009), pp. 95–114
314. A. Mohamed, A. Hasan, K. Sirén, Fulfillment of net-zero energy building (NZEB) with four metrics in a single family house with different heating alternatives. Appl. Energy **114**, 385–399 (2014)
315. US Department of Energy, Heat pump systems (2018). https://www.energy.gov/energysaver/heat-and-cool/heat-pump-systems
316. M. Electric, Mitsubishi electric wi-fi heat pump control (2018). http://www.mitsubishi-electric.co.nz/wifi/
317. Energy Information Agency EIA, Manufacturing energy consumption survey (MECS) (2013). https://www.eia.gov/consumption/manufacturing/
318. Office of Energy Efficiency and Renewable Energy (EERE-DOE), Dynamic manufacturing energy sankey tool (2010, units: Trillion btu) (2010). https://www.energy.gov/eere/amo/dynamic-manufacturing-energy-sankey-tool-2010-units-trillion-btu
319. F. Mccaffrey, Daniel Sperling of the University of California, Davis, discusses the major transformations in store for mobility (2018). http://energy.mit.edu/news/3-questions-future-transportation-systems/
320. N. Lu, N. Cheng, N. Zhang, X. Shen, J.W. Mark, Connected vehicles: solutions and challenges. IEEE Internet Things J. **1**(4), 289–299 (2014)
321. K.-M. Lee, K.I. Goh, I.M. Kim, Sandpiles on multiplex networks. J. Korean Phys. Soc. **60**(4), 641–647 (2012). http://dx.doi.org/10.3938/jkps.60.641
322. M. Amadeo, C. Campolo, A. Molinaro, Information-centric networking for connected vehicles: a survey and future perspectives. IEEE Commun. Mag. **54**(2), 98–104 (2016)
323. E. Uhlemann, Introducing connected vehicles [connected vehicles]. IEEE Veh. Technol. Mag. **10**(1), 23–31 (2015)
324. K. Adachi, N. Endo, H. Miyakoshi, Ai-based adaptive vehicle control system. US Patent 5,189,619, 23 Feb 1993
325. P.I. Labuhn, W.J. Chundrlik Jr, Adaptive cruise control. US Patent 5,454,442, 3 Oct 1995
326. H. Winner, M. Schopper, Adaptive cruise control, in *Handbook of Driver Assistance Systems: Basic Information, Components and Systems for Active Safety and Comfort* (Springer, Berlin, 2014), pp. 1–44
327. R. Rajamani, Adaptive cruise control, in *Encyclopedia of Systems and Control* (2015), pp. 13–19
328. S.E. Shladover, C.A. Desoer, J.K. Hedrick, M. Tomizuka, J. Walrand, W.-B. Zhang, D.H. McMahon, H. Peng, S. Sheikholeslam, N. McKeown, Automated vehicle control developments in the path program. IEEE Trans. Veh. Technol. **40**(1), 114–130 (1991)
329. M. Gouy, K. Wiedemann, A. Stevens, G. Brunett, N. Reed, Driving next to automated vehicle platoons: How do short time headways influence non-platoon drivers' longitudinal control? Transport. Res. F: Traffic Psychol. Behav. **27**, 264–273 (2014)
330. J. García-Nieto, E. Alba, A.C. Olivera, Swarm intelligence for traffic light scheduling: application to real urban areas. Eng. Appl. Artif. Intell. **25**(2), 274–283 (2012)
331. T. Tielert, M. Killat, H. Hartenstein, R. Luz, S. Hausberger, T. Benz, The impact of traffic-light-to-vehicle communication on fuel consumption and emissions, in *2010 Internet of Things (IOT)* (IEEE, Piscataway, 2010), pp. 1–8

332. K.E. Stoffers, Scheduling of traffic lights—a new approach. Transp. Res. **2**(3), 199–234 (1967)
333. C. Gershenson, Self-organizing traffic lights. arXiv Preprint nlin/0411066 (2004)
334. M. Khalid, Intelligent traffic lights control by fuzzy logic. Malaysian J. Comput. Sci. **9**(2), 29–35 (1996)
335. S. Arnon, Optimised optical wireless car-to-traffic-light communication. Trans. Emerg. Telecommun. Technol. **25**(6), 660–665 (2014)
336. M. Sivak, B. Schoettle, *Road Safety with Self-Driving Vehicles: General Limitations and Road Sharing with Conventional Vehicles* (Transportation Research Institute (UMTRI), Michigan, 2015)
337. N.A. Stanton, M.S. Young, Vehicle automation and driving performance. Ergonomics **41**(7), 1014–1028 (1998)
338. F.O. Flemisch, C.A. Adams, S.R. Conway, K.H. Goodrich, M.T. Palmer, P.C. Schutte, The H-Metaphor as a Guideline for Vehicle Automation and Interaction. NASA, Tech. Rep., 2003
339. M.S. Young, N.A. Stanton, What's skill got to do with it? Vehicle automation and driver mental workload. Ergonomics **50**(8), 1324–1339 (2007)
340. T. Litman, *Autonomous Vehicle Implementation Predictions* (Victoria Transport Policy Institute, Victoria, 2017)
341. F. Litvin, Z. Yi, Robotic bevel-gear differential train. Int. J. Robot. Res. **5**(2), 75–81 (1986)
342. S. Greengard, Making automation work. Commun. ACM **52**(12), 18–19 (2009)
343. V.C. Storey, H. Ullrich, S. Sundaresan, An ontology for database design automation, in *International Conference on Conceptual Modeling* (Springer, Berlin, 1997), pp. 2–15
344. H. Taso, J.L. Botha, Definition and Evaluation of Bus and Truck Automation Operations Concepts. University of California Research Reports, Tech. Rep., 2003. https://cloudfront.escholarship.org/dist/prd/content/qt9pz7n1gr/qt9pz7n1gr.pdf
345. D. Anair, A. Mahmassani, State of Charge: Electric Vehicles' Global Warming Emissions and Fuel-Cost Savings across the United States. Union of Concerned Scientists: Citizens and Scientists for Environmental Solutions, Cambridge, Tech. Rep., 2012
346. O. Karabasoglu, J. Michalek, Influence of driving patterns on life cycle cost and emissions of hybrid and plug-in electric vehicle powertrains. Energy Policy **60**, 445–461 (2013). http://www.sciencedirect.com/science/article/pii/S0301421513002255
347. L. Raykin, M.J. Roorda, H.L. MacLean, Impacts of driving patterns on tank-to-wheel energy use of plug-in hybrid electric vehicles. Transp. Res. Part D: Transp. Environ. **17**(3), 243–250 (2012). http://www.sciencedirect.com/science/article/pii/S1361920911001568
348. Z. Yang, Y. Wu, Projection of automobile energy consumption and CO_2 emissions with different propulsion/fuel system scenarios in Beijing, in *2012 2nd International Conference on Remote Sensing, Environment and Transportation Engineering (RSETE)* (IEEE, Nanjing, 2012), pp. 2012–2015
349. S. Soylu, *Electric Vehicles – The Benefits and Barriers* (Intech Open Access Publisher, Rijeka, 2011)
350. F. Carlsson, O. Johansson-Stenman, Costs and benefits of electric vehicles: a 2010 perspective. J. Trans. Econ. Policy **37**(1), 1–28 (2003). http://www.jstor.org/stable/20053920
351. Z. Li, M. Ouyang, The pricing of charging for electric vehicles in China—dilemma and solution. Energy **36**(9), 5765–5778 (2011)
352. M. Weiss, M.K. Patel, M. Junginger, A. Perujo, P. Bonnel, G. van Grootveld, On the electrification of road transport—learning rates and price forecasts for hybrid-electric and battery-electric vehicles. Energy Policy **48**, 374–393 (2012)
353. W. Sierzchula, S. Bakker, K. Maat, B.V. Wee, Environmental innovation and societal transitions the competitive environment of electric vehicles: an analysis of prototype and production models. Environ. Innov. Soc. Trans. **2**, 49–65 (2012). http://dx.doi.org/10.1016/j.eist.2012.01.004
354. A.A. Pesaran, T. Markel, H.S. Tataria, D. Howell, Battery requirements for plug-in hybrid electric vehicles – analysis and rationale, in *23rd International Electric Vehicle Symposium*, Anaheim (2007), pp. 1–18

355. M.A. Kromer, J.B. Heywood, Electric Powertrains: Opportunities and Challenges in the U.S. Light-Duty Vehicle Fleet. Massachuseets Institute of Technology, Cambridge, Tech. Rep., 2007
356. S. Manzetti, F. Mariasiu, Electric vehicle battery technologies: from resent state to future systems. Renew. Sust. Energ. Rev. **51**, 1004–1012 (2015)
357. S. Lukic, J. Cao, R. Bansal, F. Rodriguez, A. Emadi, Energy storage systems for automotive applications. IEEE Trans. Ind. Electron. **55**(6), 2258–2267 (2008)
358. S. Vazquez, S.M. Lukic, E. Galvan, L.G. Franquelo, J.M. Carrasco, Energy storage systems for transport and grid applications. IEEE Trans. Ind. Electron. **57**(12), 3881–3895 (2010)
359. A. Fotouhi, D.J. Auger, K. Propp, S. Longo, M. Wild, A review on electric vehicle battery modelling: from lithium-ion toward lithium–sulphur. Renew. Sust. Energ. Rev. **56**, 1008–1021 (2016)
360. J. Deutch, E. Moniz, Electrification of the Transportation System. Massachusetts Institute of Technology, Cambridge, Tech. Rep., 2010
361. T. Markel, Plug-in electric vehicle infrastructure: a foundation for electrified transportation, in *MIT Energy Initiative Transportation Electrification Symposium*, Cambridge (2010), pp. 1–10
362. W. Su, H. Rahimi-eichi, W. Zeng, M.-Y. Chow, A survey on the electrification of transportation in a smart grid environment. IEEE Trans. Ind. Inf. **8**(1), 1–10 (2012)
363. M.A. Hadhrami, A. Viswanath, R.A. Junaibi, A.M. Farid, S. Sgouridis, Evaluation of Electric Vehicle Adoption Potential in Abu Dhabi. Masdar Institute of Science and Technology, Abu Dhabi, Tech. Rep., 2013
364. J. Zheng, S. Mehndiratta, J.Y. Guo, Z. Liu, Strategic policies and demonstration program of electric vehicle in China. Trans. Policy **19**(1), 17–25 (2012)
365. S. Skippon, M. Garwood, Responses to battery electric vehicles: UK consumer attitudes and attributions of symbolic meaning following direct experience to reduce psychological distance. Transp. Res. Part D: Transp. Environ. **16**(7), 525–531 (2011)
366. M. Åhman, Government policy and the development of electric vehicles in Japan. Energy Policy **34**(4), 433–443 (2006)
367. B. Van Bree, G. Verbong, G. Kramer, A multi-level perspective on the introduction of hydrogen and battery-electric vehicles. Technol. Forecast. Soc. Chang. **77**(4), 529–540 (2010). http://linkinghub.elsevier.com/retrieve/pii/S004016250900211X
368. J.A. Schellenberg, M.J.S. Member, Electric vehicle forecast for a large west coast utility, in *2011 IEEE Power and Energy Society General Meeting* (IEEE, San Diego, 2011), pp. 1–6
369. K. Yabe, Y. Shinoda, T. Seki, H. Tanaka, A. Akisawa, Market penetration speed and effects on CO_2 reduction of electric vehicles and plug-in hybrid electric vehicles in Japan. Energy Policy **45**, 529–540 (2012). http://dx.doi.org/10.1016/j.enpol.2012.02.068
370. A. Burke, Present status and marketing prospects of the emerging hybrid-electric and diesel technologies to reduce CO_2 emissions of new light-duty vehicles in California. Contract **2**, 1–24 (2004)
371. K. Parks, P. Denholm, T. Markel, Costs and Emissions Associated with Plug-in Hybrid Electric Vehicle Charging in the Xcel Energy Colorado Service Territory. National Renewable Energy Laboratory, Golden, CO, Tech. Rep., 2007
372. K. Sikes, T. Gross, Z. Lin, J. Sullivan, T. Cleary, J. Ward, Plug-in Hybrid Electric Vehicle Market Introduction Study Final Report. Sentech Inc., Washington, Tech. Rep., 2010
373. EPRI, *Environmental Assessment of Plug-in Hybrid Electric Vehicles* (Electric Power Research Institute, Palo Alto, 2007)
374. Y. Zhou, P. Mancarella, J. Mutale, Modelling and assessment of the contribution of demand response and electrical energy storage to adequacy of supply. Sustain. Energy Grids Netw. **3**, 2–23 (2015)
375. OCED-IEA-Anonymous, A.L. At, T.H.E. Global, E. Vehicle, Ev City Casebook: A Look at the Global Electric Vehicle Movement. Organisation for Economic Cooperation and Development/International Energy Agency, Paris, Tech. Rep., 2012
376. U. Tietge, P. Mock, N. Lutsey, A. Campestrini, Comparison of leading electric vehicle policy and deployment in Europe. Communications **49**(30), 847129-102 (2016)

377. J. Pointon, The multi-unit dwelling vehicle charging challenge, in *Electric Vehicles Virtual Summit 2012*, vol. 69 (The Smart Grid Observer, 2012)
378. J. Kassakian, R. Schmalensee, G. Desgroseilliers, T. Heidel, K. Afridi, A. Farid, J. Grochow, W. Hogan, H. Jacoby, J. Kirtley, H. Michaels, I. Perez-Arriaga, D. Perreault, N. Rose, G. Wilson, N. Abudaldah, M. Chen, P. Donohoo, S. Gunter, P. Kwok, V. Sakhrani, J. Wang, A. Whitaker, X. Yap, R. Zhang, MI Technology, The Future of the Electric Grid: An Interdisciplinary MIT Study, 2011
379. R. Al Junaibi, A.M. Farid, A method for the technical feasibility assessment of electric vehicle penetration, in *3rd MIT-MI Joint Workshop on the Reliability of Power System Operation and Control in the Presence of Increasing Penetration of Variable Energy Sources*. 2012, pp. 1–16
380. R. Al Junaibi, A.M. Farid, A method for the technical feasibility assessment of electrical vehicle penetration, in *7th Annual IEEE Systems Conference* (IEEE, Orlando, 2013), pp. 1–6. http://dx.doi.org/10.1109/SysCon.2013.6549945
381. A. Viswanath, A.M. Farid, A hybrid dynamic system model for the assessment of transportation electrification, in *2014 American Control Conference* (IEEE, Portland, 2014), pp. 1–7. http://dx.doi.org/10.1109/ACC.2014.6858810
382. A. Viswanath, E.E.S. Baca, A.M. Farid, An axiomatic design approach to passenger itinerary enumeration in reconfigurable transportation systems. IEEE Trans. Intell. Transp. Syst. **15**(3), 915–924 (2014). http://dx.doi.org/10.1109/TITS.2013.2293340
383. E.E.S. Baca, A.M. Farid, An axiomatic design approach to reconfigurable transportation systems planning and operations (invited paper), in *DCEE 2013: 2nd International Workshop on Design in Civil and Environmental Engineering* (Mary Kathryn Thompson, Worcester, 2013), pp. 22–29. http://engineering.dartmouth.edu/liines/resources/Conferences/TES-C11.pdf
384. A.M. Farid, I.S. Khayal, Axiomatic design based volatility assessment of the Abu Dhabi healthcare labor market. Part I: theory, in *Proceedings of the ICAD 2013: The Seventh International Conference on Axiomatic Design* (Worcester, 2013), pp. 1–8. http://engineering.dartmouth.edu/liines/resources/Conferences/IES-C21.pdf
385. A.M. Farid, Static resilience of large flexible engineering systems. Part I – axiomatic design model, in *4th International Engineering Systems Symposium* (Stevens Institute of Technology, Hoboken, 2014), pp. 1–8. http://engineering.dartmouth.edu/liines/resources/Conferences/IES-C37.pdf
386. A.M. Farid, Static resilience of large flexible engineering systems. Part II – axiomatic design measures, in *4th International Engineering Systems Symposium* (Stevens Institute of Technology, Hoboken, 2014), pp. 1–8. http://engineering.dartmouth.edu/liines/resources/Conferences/IES-C38.pdf
387. A.M. Farid, Static resilience of large flexible engineering systems: axiomatic design model and measures. IEEE Syst. J. **PP**(99), 1–12 (2015). http://dx.doi.org/10.1109/JSYST.2015.2428284
388. R. Al Junaibi, A.M. Farid, Technical feasibility assessment of Abu Dhabi ITS for electric vehicles, in *2nd MHI-MI Joint Workshop on the Electric Vehicle Adoption Feasibility in Abu Dhabi* (Masdar Institute of Science and Technology, Abu Dhabi, 2013), pp. 1–27
389. R. Al Junaibi, A. Viswanath, A.M. Farid, Technical feasibility assessment of electric vehicles: an Abu Dhabi example, in *2nd IEEE International Conference on Connected Vehicles and Expo* (IEEE, Las Vegas, 2013), pp. 410–417. http://dx.doi.org/10.1109/ICCVE.2013.6799828
390. M.A. Hadhrami, A. Viswanath, R. Al Junaibi, A.M. Farid, S. Sgouridis, Evaluation of Electric Vehicle Adoption Potential in Abu Dhabi [ETN-W02]. Masdar Institute of Science and Technology, Abu Dhabi, Tech. Rep., 2013
391. T. Sonoda, K. Kawaguchi, Y. Kamino, Y. Koyanagi, H. Ogawa, H. Ono, Environment-conscious urban design simulator "clean mobility simulator"—traffic simulator that includes electric vehicles. Mitsubishi Heavy Ind. Tech. Rev. **49**(1), 78–83 (2012)
392. D.F. Allan, A.M. Farid, A benchmark analysis of open source transportation-electrification simulation tools, in *2015 IEEE Conference on Intelligent Transportation Systems* (IEEE, Las Palmas, 2015), pp. 1–7. http://dx.doi.org/10.1109/ITSC.2015.198

393. N. Agatz, A. Erera, M. Savelsbergh, X. Wang, Optimization for dynamic ride-sharing: a review. Eur. J. Oper. Res. **223**(2), 295–303 (2012)
394. J. Alonso-Mora, S. Samaranayake, A. Wallar, E. Frazzoli, D. Rus, On-demand high-capacity ride-sharing via dynamic trip-vehicle assignment. Proc. Natl. Acad. Sci. **114**(3), 462–467 (2017)
395. N. Agatz, A.L. Erera, M.W. Savelsbergh, X. Wang, Dynamic ride-sharing: a simulation study in metro Atlanta. Procedia. Soc. Behav. Sci. **17**, 532–550 (2011)
396. L. Pieltain Fernandez, T. Gomez San Roman, R. Cossent, C. Mateo Domingo, P. Frias, Assessment of the impact of plug-in electric vehicles on distribution networks. IEEE Trans. Power Syst. **26**(1), 206–213 (2011)
397. J.A.P. Lopes, F.J. Soares, P.M.R. Almeida, Integration of electric vehicles in the electric power system. Proc. IEEE **99**(1), 168–183 (2011)
398. K. Qian, C. Zhou, M. Allan, Y. Yuan, Modeling of load demand due to EV battery charging in distribution systems. IEEE Trans. Power Syst. **26**(2), 802–810 (2011)
399. K.J. Dyke, N. Schofield, M. Barnes, The impact of transport electrification on electrical networks. IEEE Trans. Ind. Electron. **57**(12), 3917–3926 (2010)
400. A.Y. Saber, G.K. Venayagamoorthy, Plug-in vehicles and renewable energy sources for cost and emission reductions. IEEE Trans. Ind. Electron. **58**(4), 1229–1238 (2011)
401. L. Gan, U. Topcu, S. Low, Optimal decentralized protocol for electric vehicle charging. IEEE Trans. Power Syst. **28**(2), 940–951 (2013)
402. M. Erol-Kantarci, J.H. Sarker, H.T. Mouftah, Quality of service in plug-in electric vehicle charging infrastructure, in *2012 IEEE International Electric Vehicle Conference (IEVC)* (IEEE, Greenville, 2012), pp. 1–5
403. Z. Ma, D. Callaway, I. Hiskens, Optimal charging control for plug-in electric vehicles, in *Control and Optimization Methods for Electric Smart Grids* (Springer, Berlin, 2012), pp. 259–273
404. Q. Gong, S. Midlam-Mohler, E. Serra, V. Marano, G. Rizzoni, PEV charging control for a parking lot based on queuing theory, in *2013 American Control Conference* (IEEE, Washington, 2013), pp. 1126–1131
405. W. Kempton, J. Tomić, Vehicle-to-grid power implementation: from stabilizing the grid to supporting large-scale renewable energy. J. Power Sources **144**(1), 280–294 (2005)
406. W.C. Schoonenberg, A.M. Farid, Modeling smart cities with hetero-functional graph theory, in *2017 IEEE International Conference on Systems, Man, and Cybernetics (SMC2017), Intelligent Industrial System Special Session*, vol. 1 (IEEE, Piscataway, 2017), pp. 1–10
407. W.C. Schoonenberg, I.S. Khayal, A.M. Farid, *A Hetero-functional Graph Theory for Modeling Smart City Infrastructure* (Springer, Berlin, 2018)
408. M. Yigit, V.C. Gungor, G. Tuna, M. Rangoussi, E. Fadel, Power line communication technologies for smart grid applications: a review of advances and challenges. Comput. Netw. **70**, 366–383 (2014)
409. K. Sharma, L.M. Saini, Power-line communications for smart grid: progress, challenges, opportunities and status. Renew. Sust. Energ. Rev. **67**, 704–751 (2017)
410. M. Kuzlu, M. Pipattanasomporn, S. Rahman, Communication network requirements for major smart grid applications in HAN, NAN and WAN. Comput. Netw. **67**, 74–88 (2014)
411. D.J. Marihart, Communications technology guidelines for EMS/SCADA systems. IEEE Trans. Power Delivery **16**(2), 181–188 (2001)
412. H. Chen, W.J. Zheng, Q.Z. Meng, C.Q. Zhao, Q.D. Li, X.J. Lv, A survey of communication technology in distribution network, in *IEEE PES Innovative Smart Grid Technologies* (IEEE, Piscataway, 2012), pp. 1–6
413. P.P. Tsang, S.W. Smith, YASIR: a low-latency, high-integrity security retrofit for legacy SCADA systems, in *IFIP International Information Security Conference* (Springer, Berlin, 2008), pp. 445–459
414. R.H. Khan, J.Y. Khan, A comprehensive review of the application characteristics and traffic requirements of a smart grid communications network. Comput. Netw. **57**(3), 825–845 (2013)

415. WiSUNAlliance, Comparing IoT Networks at a Glance: How Wi-SUN Compares with LoRaWAN and NB-IoT. Wi-SUN Alliance, Tech. Rep., 2016
416. J. Petajajarvi, K. Mikhaylov, A. Roivainen, T. Hanninen, M. Pettissalo, On the coverage of LPWANs: range evaluation and channel attenuation model for LoRa technology, in *2015 14th International Conference on ITS Telecommunications (ITST)* (IEEE, Piscataway, 2015), pp. 55–59
417. F. Adelantado, X. Vilajosana, P. Tuset-Peiro, B. Martinez, J. Melia-Segui, T. Watteyne, Understanding the limits of LoRaWAN. IEEE Commun. Mag. **55**(9), 34–40 (2017)
418. M. Centenaro, L. Vangelista, A. Zanella, M. Zorzi, Long-range communications in unlicensed bands: the rising stars in the IoT and smart city scenarios. IEEE Wireless Commun. **23**(5), 60–67 (2016)
419. What is sigfox? (2015). https://www.link-labs.com/blog/what-is-sigfox
420. SigFox vs. LoRa: a comparison between technologies and business models (2018). https://www.link-labs.com/blog/sigfox-vs-lora
421. A.D. Zayas, P. Merino, The 3GPP NB-IoT system architecture for the Internet of Things, in *2017 IEEE International Conference on Communications Workshops (ICC Workshops)* (IEEE, Piscataway, 2017), pp. 277–282
422. M. Chen, Y. Miao, Y. Hao, K. Hwang, Narrow band internet of things. IEEE Access **5**, 20557–20577 (2017)
423. A. Sabbah, A. El-Mougy, M. Ibnkahla, A survey of networking challenges and routing protocols in smart grids. IEEE Trans. Ind. Inf. **10**(1), 210–221 (2014)
424. A. Mahmood, N. Javaid, S. Razzaq, A review of wireless communications for smart grid. Renew. Sustain. Energy Rev. **41**, 248–260 (2015)
425. C.-X. Wang, F. Haider, X. Gao, X.-H. You, Y. Yang, D. Yuan, H. Aggoune, H. Haas, S. Fletcher, E. Hepsaydir, Cellular architecture and key technologies for 5G wireless communication networks. IEEE Commun. Mag. **52**(2), 122–130 (2014)
426. A. Kailas, V. Cecchi, A. Mukherjee, A survey of communications and networking technologies for energy management in buildings and home automation. Int. J. Comput. Netw. Commun. Secur. **2012**, 932181 (2012)
427. M. Shahraeini, M. Javidi, M. Ghazizadeh, A new approach for classification of data transmission media in power systems, in *2010 International Conference on Power System Technology (POWERCON)* (IEEE, Piscataway, 2010), pp. 1–7
428. C.H. Hauser, D.E. Bakken, A. Bose, A failure to communicate: next generation communication requirements, technologies, and architecture for the electric power grid. IEEE Power Energ. Mag. **3**(2), 47–55 (2005). http://dx.doi.org/10.1109/MPAE.2005.1405870
429. C. Lavenu, D. Dufresne, X. Montuelle, Innovative solution sustaining supervisory control and data acquisition to remote terminal unit G3-PLC connectivity over dynamic grid topologies. CIRED-Open Access Proc. J. **2017**(1), 1237–1241 (2017)
430. G.N. Ericsson, Cyber security and power system communicationessential parts of a smart grid infrastructure. IEEE Trans. Power Delivery **25**(3), 1501–1507 (2010)
431. H. Khurana, M. Hadley, N. Lu, D. Frincke, Smart-grid security issues. IEEE Secur. Priv. **8**(1), 81–85 (2010)
432. R.S. Sinha, Y. Wei, S.-H. Hwang, A survey on LPWA technology: LoRa and NB-IoT. ICT Express **3**(1), 14–21 (2017)
433. J.-P. Bardyn, T. Melly, O. Seller, N. Sornin, IoT: the era of LPWAN is starting now, in *ESSCIRC Conference 2016: 42nd European Solid-State Circuits Conference* (IEEE, Piscataway, 2016), pp. 25–30
434. N. Sornin, M. Luis, T. Eirich, T. Kramp, O. Hersent, *Lorawan Specification* (LoRa Alliance, 2015)
435. Background information about LoRaWAN (2018). https://www.thethingsnetwork.org/docs/lorawan/
436. C. McClelland, IoT connectivity – comparing NB-IoT, LTE-M, LoRa, SigFox, and other LPWAN technologies (2018). https://www.iotforall.com/iot-connectivity-comparison-lora-sigfox-rpma-lpwan-technologies/

437. SEMTECH, Radiation Leak Detection. SEMTECH Corporation, Tech. Rep., 2017. https://www.semtech.com/uploads/technology/LoRa/app-briefs/Semtech_Enviro_RadiationLeak_AppBrief-FINAL.pdf
438. SEMTECH, Air Pollution Monitoring. SEMTECH Corporation, Tech. Rep., 2017. https://www.semtech.com/uploads/technology/LoRa/app-briefs/Semtech_Enviro_AirPollution_AppBrief-FINAL.pdf
439. R. Ratasuk, B. Vejlgaard, N. Mangalvedhe, A. Ghosh, NB-IoT system for M2M communication, in *2016 IEEEWireless Communications and Networking Conference (WCNC)* (IEEE, Piscataway, 2016), pp. 1–5
440. HUAWEI, NB-IoT Enabling New Business Opportunities. HUAWEI Technologies Co., Tech. Rep., 2015
441. HUAWEI, NB-IoT Commercial Premier Use Case Library. HUAWEI Technologies Co., Tech. Rep., 2017. https://www.gsma.com/iot/wp-content/uploads/2017/12/NB-IoT-Commercial-Premier-Use-case-Library-1.0_Layout_171110.pdf
442. https://www.sigfox.com/en/utilities-energy (2018)
443. WiSUNAlliance, Wisun Alliance (2017). https://www.wi-sun.org/images/assets/docs/Wi-SUN_Organization-v1.pdf
444. Wi-SUN Alliance Field Area Network (FAN) Overview (2016). https://datatracker.ietf.org/meeting/97/materials/slides-97-lpwan-35-wi-sun-presentation-00
445. K.-H. Chang, B. Mason, The IEEE 802.15. 4G standard for smart metering utility networks, in *2012 IEEE Third International Conference on Smart Grid Communications (SmartGridComm)* (IEEE, Piscataway, 2012), pp. 476–480
446. B. Liu, M. Zhang, WiSUN Use Cases. HUAWEI Technologies Co., Tech. Rep., 2017
447. M. Arenas-Martinez, S. Herrero-Lopez, A. Sanchez, J.R. Williams, P. Roth, P. Hofmann, A. Zeier, A comparative study of data storage and processing architectures for the smart grid, in *2010 First IEEE International Conference on Smart Grid Communications (SmartGridComm)* (IEEE, Piscataway, 2010), pp. 285–290
448. Y. Yang, H. Hu, J. Xu, G. Mao, Relay technologies for WiMax and LTE-advanced mobile systems. IEEE Commun. Mag. **47**(10), 100–105 (2009)
449. A. Sarkar, J. Pick, G. Moss, Geographic patterns and socio-economic influences on mobile internet access and use in united states counties, in *Proceedings of the 50th Hawaii International Conference on System Sciences* (2017)
450. K. Christensen, P. Reviriego, B. Nordman, M. Bennett, M. Mostowfi, J.A. Maestro, IEEE 802.3az: the road to energy efficient ethernet. IEEE Commun. Mag. **48**(11), 50–56 (2010)
451. T. Salonidis, P. Bhagwat, L. Tassiulas, Proximity awareness and fast connection establishment in bluetooth, in *Proceedings of the 1st ACM International Symposium on Mobile Ad Hoc Networking and Computing* (IEEE Press, Boston, 2000), pp. 141–142
452. K. Sairam, N. Gunasekaran, S.R. Redd, Bluetooth in wireless communication. IEEE Commun. Mag. **40**(6), 90–96 (2002)
453. I. Ungurean, N.-C. Gaitan, V.G. Gaitan, An IoT architecture for things from industrial environment, in *2014 10th International Conference on Communications (COMM)* (IEEE, Piscataway, 2014), pp. 1–4
454. Z. Sheng, C. Mahapatra, C. Zhu, V.C. Leung, Recent advances in industrial wireless sensor networks toward efficient management in IoT. IEEE access **3**, 622–637 (2015)
455. M. Wollschlaeger, T. Sauter, J. Jasperneite, The future of industrial communication: automation networks in the era of the internet of things and industry 4.0. IEEE Ind. Electron. Mag. **11**(1), 17–27 (2017)
456. X. Li, D. Li, J. Wan, A.V. Vasilakos, C.-F. Lai, S. Wang, A review of industrial wireless networks in the context of industry 4.0. Wirel. Netw **23**(1), 23–41 (2017)
457. S. Jeschke, C. Brecher, T. Meisen, D. Özdemir, T. Eschert, Industrial internet of things and cyber manufacturing systems, in *Industrial Internet of Things* (Springer, Berlin, 2017), pp. 3–19
458. M.N. Sadiku, Y. Wang, S. Cui, S.M. Musa, Industrial internet of things. IJASRE **3**(11), 1–5 (2017)

459. D. Serpanos, M. Wolf, Industrial internet of things, in *Internet-of-Things (IoT) Systems* (Springer, Berlin, 2018), pp. 37–54
460. C.-Y. Chong, S.P. Kumar, Sensor networks: evolution, opportunities, and challenges. Proc. IEEE **91**(8), 1247–1256 (2003)
461. A. Hakiri, P. Berthou, A. Gokhale, S. Abdellatif, Publish/subscribe-enabled software defined networking for efficient and scalable IoT communications. IEEE Commun. Mag. **53**(9), 48–54 (2015)
462. A. Al-Fuqaha, A. Khreishah, M. Guizani, A. Rayes, M. Mohammadi, Toward better horizontal integration among IoT services. IEEE Commun. Mag. **53**(9), 72–79 (2015)
463. DDS Standard (2017–2018). https://www.rti.com/products/dds-standard
464. V. Karagiannis, P. Chatzimisios, F. Vazquez-Gallego, J. Alonso-Zarate, A survey on application layer protocols for the internet of things. Trans. IoT Cloud Comput. **3**(1), 11–17 (2015)
465. J. Lee, B. Park, Development and evaluation of a cooperative vehicle intersection control algorithm under the connected vehicles environment. IEEE Trans. Intell. Transp. Syst. **13**(1), 81–90 (2012)
466. D. Thangavel, X. Ma, A. Valera, H.-X. Tan, C.K.-Y. Tan, Performance evaluation of MQTT and CoAP via a common middleware, in *2014 IEEE Ninth International Conference on Intelligent Sensors, Sensor Networks and Information Processing (ISSNIP)* (IEEE, Piscataway, 2014), pp. 1–6
467. A.P. Castellani, M. Gheda, N. Bui, M. Rossi, M. Zorzi, Web services for the internet of things through CoAP and EXI, in *2011 IEEE International Conference on Communications Workshops (ICC)* (IEEE, Piscataway, 2011), pp. 1–6
468. M.R. Palattella, N. Accettura, X. Vilajosana, T. Watteyne, L.A. Grieco, G. Boggia, M. Dohler, Standardized protocol stack for the internet of (important) things. IEEE Commun. Surv. Tutorials **15**(3), 1389–1406 (2013)
469. S.J. Vaughan-Nichols, Google moves away from the XMPP open-messaging standard (2013). GooglemovesawayfromtheXMPPopen-messagingstandard
470. S. Bendel, T. Springer, D. Schuster, A. Schill, R. Ackermann, M. Ameling, A service infrastructure for the internet of things based on XMPP, in *2013 IEEE International Conference on Pervasive Computing and Communications Workshops (PERCOM Workshops)* (IEEE, Piscataway, 2013), pp. 385–388
471. F.T. Johnsen, T.H. Bloebaum, M. Avlesen, S. Spjelkavik, B. Vik, Evaluation of transport protocols for web services, in *2013 Military Communications and Information Systems Conference (MCC)* (IEEE, Piscataway, 2013), pp. 1–6
472. S.O. Johnsen, M. Veen, Risk assessment and resilience of critical communication infrastructure in railways. Cogn. Tech. Work **15**(1), 95–107. http://dx.doi.org/10.1007/s10111-011-0187-2
473. W. Saad, Z. Han, H.V. Poor, T. Basar, Game-theoretic methods for the smart grid: an overview of microgrid systems, demand-side management, and smart grid communications. IEEE Signal Process. Mag. **29**(5), 86–105 (2012)
474. A. Muzhikyan, A.M. Farid, K.Youcef-Toumi, An enhanced method for the determination of the regulation reserves, in *IEEE American Control Conference* (IEEE, Los Angeles, 2015), pp. 1–8. http://dx.doi.org/10.1109/ACC.2015.7170866
475. A. Muzhikyan, A.M. Farid, K. Youcef-Toumi, An enhanced method for the determination of the ramping reserves, in *IEEE American Control Conference* (IEEE, Los Angeles, 2015), pp. 1–8. http://dx.doi.org/10.1109/ACC.2015.7170863
476. A. Muzhikyan, A.M. Farid, K. Youcef-Toumi, An enhanced method for the determination of load following reserves, in *2014 American Control Conference* (IEEE, Portland, 2014), pp. 1–8. http://dx.doi.org/10.1109/ACC.2014.6859254
477. A. Muzhikyan, A.M. Farid, K. Youcef-Toumi, A power grid enterprise control method for energy storage system integration, in *IEEE Innovative Smart Grid Technologies Conference Europe*, Istanbul, 2014, pp. 1–6. http://dx.doi.org/10.1109/ISGTEurope.2014.7028898
478. A. Muzhikyan, T. Mezher, A.M. Farid, Power system enterprise control with inertial response procurement. IEEE Trans. Power Syst. **33**(4), 3735–3744 (2018)

479. P. Vrba, V. Marik, P. Siano, P. Leitao, G. Zhabelova, V. Vyatkin, T. Strasser, A review of agent and service-oriented concepts applied to intelligent energy systems. IEEE Trans. Ind. Inf. **10**(3), 1890–1903 (2014)
480. V.C. Güngör, D. Sahin, T. Kocak, S. Ergut, C. Buccella, C. Cecati, G.P. Hancke, Smart grid and smart homes: key players and pilot projects. IEEE Ind. Electron. Mag. **6**(4), 18–34 (2012)
481. W. Lu, M. Liu, S. Lin, L. Li, Fully decentralized optimal power flow of multi-area interconnected power systems based on distributed interior point method. IEEE Trans. Power Syst. **33**(1), 901–910 (2018)
482. A.R. Soares, O. De Somer, D. Ectors, F. Aben, J. Goyvaerts, M. Broekmans, F. Spiessens, D. van Goch, K. Vanthournout, Distributed optimization algorithm for residential flexibility activation-results from a field test. IEEE Trans. Power Syst. (2018). http://dx.doi.org/10.1109/TPWRS.2018.2809440
483. B. Houska, J. Frasch, M. Diehl, An augmented lagrangian based algorithm for distributed nonconvex optimization. SIAM J. Optim. **26**(2), 1101–1127 (2016)
484. M. Hong, Z.-Q. Luo, M. Razaviyayn, Convergence analysis of alternating direction method of multipliers for a family of nonconvex problems. SIAM J. Optim. **26**(1), 337–364 (2016)
485. A. Engelmann, T. Mühlpfordt, Y. Jiang, B. Houska, T. Faulwasser, Distributed AC optimal power flow using ALADIN. IFAC-PapersOnLine **50**(1), 5536–5541 (2017)
486. D. Hur, J.-K. Park, B.H. Kim, Evaluation of convergence rate in the auxiliary problem principle for distributed optimal power flow. IEE Proc. Gener. Transm. Distrib. **149**(5), 525–532 (2002)
487. J.-H. Hours, C.N. Jones, An alternating trust region algorithm for distributed linearly constrained nonlinear programs. J. Optim. Theory Appl. **173**(3), 844–877 (2017)
488. J. Guo, G. Hug, O.K. Tonguz, A case for non-convex distributed optimization in large-scale power systems. IEEE Trans. Power Syst. **32**(5), 3842–3851 (2017)
489. J. Zhang, S. Nabavi, A. Chakrabortty, Y. Xin, Convergence analysis of ADMM-based power system mode estimation under asynchronous wide-area communication delays, in *2015 IEEE Power & Energy Society General Meeting* (IEEE, Piscataway, 2015)
490. X. Bai, H. Wei, K. Fujisawa, Y. Wang, Semidefinite programming for optimal power flow problems. Int. J. Electr. Power Energy Syst. **30**(6–7), 383–392 (2008)
491. D.K. Molzahn, J.T. Holzer, B.C. Lesieutre, C.L. DeMarco, Implementation of a large-scale optimal power flow solver based on semidefinite programming. IEEE Trans. Power Syst. **28**(4), 3987–3998 (2013)
492. Q. Peng, S.H. Low, Distributed optimal power flow algorithm for radial networks, I: balanced single phase case. IEEE Trans. Smart Grid **9**(1), 111–121 (2018)
493. S.H. Low, Convex relaxation of optimal power flow – Part II: exactness. IEEE Trans. Control Netw. Syst. **1**(2), 177–189 (2014)
494. T. Erseghe, A distributed approach to the OPF problem. EURASIP J. Adv. Signal Process. **2015**, 45 (2015)
495. B.H. Kim, R. Baldick, Coarse-grained distributed optimal power flow. IEEE Trans. Power Syst. **12**(2), 932–939 (1997)
496. S.H. Low, Convex relaxation of optimal power flow – Part I: formulations and equivalence. IEEE Trans. Control Netw. Syst. **1**(1), 15–27 (2014)
497. M. Arnold, S. Knopfli, G. Andersson, Improvement of OPF decomposition methods applied to multi-area power systems, in *Power Tech, 2007 IEEE Lausanne* (IEEE, Lausanne, 2007), pp. 1308–1313
498. K. Christakou, D.-C. Tomozei, J.-Y. Le Boudec, M. Paolone, AC OPF in radial distribution networks – Part I: on the limits of the branch flow convexification and the alternating direction method of multipliers. Electr. Power Syst. Res. **143**, 438–450 (2017)
499. J. Lavaei, S.H. Low, Zero duality gap in optimal power flow problem. IEEE Trans. Power Syst. **27**(1), 92–107 (2012)
500. B.H. Kim, R. Baldick, A comparison of distributed optimal power flow algorithms. IEEE Trans. Power Syst. **15**(2), 599–604 (2000)

501. D.K. Molzahn, C. Josz, I.A. Hiskens, P. Panciatici, A Laplacian-based approach for finding near globally optimal solutions to OPF problems. IEEE Trans. Power Syst. **32**(1), 305–315 (2017)
502. M. Ilic, Reconciling hierarchical control and open access in the changing electric power industry, in *Proceedings of the 2005 IEEE Networking, Sensing and Control* (IEEE, Piscataway, 2005), pp. 799–809
503. A. Bidram, F.L. Lewis, A. Davoudi, Distributed control systems for small-scale power networks: using multiagent cooperative control theory. IEEE Control Syst. **34**(6), 56–77 (2014)
504. S. McArthur, E. Davidson, Concepts and approaches in multi-agent systems for power applications, (invited paper), in *Proceedings of the 13th International Conference on Intelligent Systems Application to Power Systems* (IEEE, Piscataway, 2005), p. 5
505. S.D.J. McArthur, E.M. Davidson, V.M. Catterson, A.L. Dimeas, N.D. Hatziargyriou, F. Ponci, T. Funabashi, Multi-agent systems for power engineering applications: Part II: technologies, standards, and tools for building multi-agent systems. IEEE Trans. Power Syst. **22**(4), 1753–1759 (2007)
506. A. Bagnall, G. Smith, A multiagent model of the UK market in electricity generation. IEEE Trans. Evol. Comput. **9**(5), 522–536 (2005)
507. H. Zhou, Z. Tu, S. Talukdar, K.C. Marshall, Wholesale electricity market failure and the new market design, in *IEEE Power Engineering Society General Meeting, 2005*, vol. 1 (IEEE, San Francisco, 2005), pp. 503–508
508. I. Praca, C. Ramos, Z. Vale, M. Cordeiro, I. Praça, Intelligent agents for negotiation and game-based decision support in electricity markets. Int. J. Eng. Intell. Syst. **13**(2), 147–154 (2005)
509. D. Sharma, D. Srinivasan, A. Trivedi, Multi-agent approach for profit based unit commitment, in *2011 IEEE Congress of Evolutionary Computation (CEC)* (IEEE, New Orleans, 2011), pp. 2527–2533
510. D. Sharma, A. Trivedi, D. Srinivasan, L. Thillainathan, Multi-agent modeling for solving profit based unit commitment problem. Appl. Soft Comput. J. **13**(8), 3751–3761 (2013). http://dx.doi.org/10.1016/j.asoc.2013.04.001
511. A. Motto, F. Galiana, Equilibrium of auction markets with unit commitment: the need for augmented pricing. IEEE Trans. Power Syst. **17**(3), 798–805 (2002)
512. A. Motto, F. Galiana, A. Conejo, M. Huneault, On walrasian equilibrium for pool-based electricity markets. IEEE Trans. Power Syst. **17**(3), 774–781 (2002)
513. A.L. Motto, F.D. Galiana, Closure on "Equilibrium of auction markets with unit commitment: the need for augmented pricing." IEEE Trans. Power Syst. **18**(2), 961–962 (2003)
514. T. Nagata, H. Sasaki, A multi-agent approach to power system restoration. IEEE Trans. Power Syst. **17**(2), 457–462 (2002)
515. T. Nagata, M. Ohono, J. Kubokawa, H. Sasaki, H. Fujita, A multi-agent approach to unit commitment problems, in *2002 IEEE Power Engineering Society Winter Meeting. Conference Proceedings (Cat. No.02CH37309)*, vol. 1 (IEEE, Piscataway, 2002), pp. 64–69
516. J.M. Zolezzi, H. Rudnick, Transmission cost allocation by cooperative games and coalition formation. IEEE Trans. Power Syst. **17**(4), 1008–1015 (2002)
517. C.S.K. Yeung, A.S.Y. Poon, F.F. Wu, Game theoretical multi-agent modelling of coalition formation for multilateral trades. IEEE Trans. Power Syst. **14**(3), 929–934 (1999)
518. P. Wei, Y. Yan, Y. Ni, J. Yen, F.F. Wu, A decentralized approach for optimal wholesale cross-border trade planning using multi-agent technology. IEEE Trans. Power Syst. **16**(4), 833–838 (2001)
519. P. Wei, Y. Yan, Y. Ni, J. Yen, F.F. Wu, A multi-agent based negotiation support system for cost allocation of cross-border transmission. Intell. Syst. Account. Finance Manag. **10**(3), 187–200 (2001)
520. P. Wei, Y. Yan, Y. Ni, J. Yen, F.F. Wu, Allocation of tie-line costs in power exchange scheduling using a multi-agent approach, in *2001 IEEE Power Engineering Society Winter Meeting. Conference Proceedings (Cat. No.01CH37194)*, vol. 3 (IEEE, Columbus, 2001), pp. 1220–1225

521. T. Kato, H. Kanamori, Y. Suzuoki, T. Funabashi, S. Member, Multi-agent based control and protection of power distribution system – protection scheme with simplified information utilization, in *Proceedings of the 13th International Conference on Intelligent Systems Application to Power Systems* (IEEE, Piscataway, 2005), pp. 49–54
522. J. Yu, J. Zhou, B. Hua, R. Liao, Optimal short-term generation scheduling with multi-agent system under a deregulated power. Int. J. Comput. Cogn. **3**(2), 61–65 (2005)
523. M. Nordman, M. Lehtonen, Distributed agent-based state estimation for electrical distribution networks. IEEE Trans. Power Syst. **20**(2), 652–658 (2005)
524. H. Wan, K.K. Li, K.P. Wong, An multi-agent approach to protection relay coordination with distributed generators in industrial power distribution system, in *Fourtieth IAS Annual Meeting. Conference Record of the 2005 Industry Applications Conference, 2005*, vol. 2 (IEEE, Kowloon, 2005), pp. 830–836
525. J.-H. Li, W.-J. Liu, Development of an agent-based system for collaborative multi-project planning and scheduling, *2005 International Conference on Machine Learning and Cybernetics*, vol. 1 (IEEE, Piscataway, 2005), pp. 119–124
526. T. Logenthiran, D. Srinivasan, T.Z. Shun, Multi-agent system for demand side management in smart grid, in *2011 IEEE Ninth International Conference on Power Electronics and Drive Systems* (IEEE, Singapore, 2011), pp. 424–429
527. Z.A. Vale, H. Morais, H. Khodr, Intelligent multi-player smart grid management considering distributed energy resources and demand response, in *IEEE PES General Meeting* (IEEE, Providence, 2010), pp. 1–7
528. M.E. Elkhatib, R. El-Shatshat, M.M.A. Salama, Novel coordinated voltage control for smart distribution networks with DG. IEEE Trans. Smart Grid **2**(4), 598–605 (2011)
529. T. Soares, H. Morais, B. Canizes, Z. Vale, Energy and ancillary services joint market simulation, in *2011 8th International Conference on the European Energy Market (EEM)* (IEEE, Zagreb, 2011), pp. 262–267
530. N.C. Ekneligoda, W.W. Weaver, Optimal transient control of microgrids using a game theoretic approach, in *2011 IEEE Energy Conversion Congress and Exposition* (IEEE, Piscataway, 2011), pp. 935–942
531. N.C. Ekneligoda, W.W. Weaver, Game-theoretic communication structures in microgrids. IEEE Trans. Power Delivery **27**(4), 2334–2341 (2012)
532. N.C. Ekneligoda, W.W. Weaver, A game theoretic bus selection method for loads in multibus DC power systems. IEEE Trans. Ind. Electron. **61**(4), 1669–1678 (2014)
533. N. Ekneligoda, W. Weaver, Game Theoretic cold-start transient optimization in DC microgrids. IEEE Trans. Ind. Electron. **61**(12), 6681–6690 (2014)
534. N.C. Ekneligoda, W.W. Weaver, Game-theoretic cold-start transient optimization in DC microgrids. IEEE Trans. Ind. Electron. **61**(12), 6681–6690 (2014).
535. N.C. Ekneligoda, W.W. Weaver, Game theoretic optimization of DC micro-grids without a communication infrastructure, in *2014 Clemson University Power Systems Conference* (IEEE, Piscataway, 2014), pp. 1–6
536. W. Saad, Z. Han, H.V. Poor, Coalitional game theory for cooperative micro-grid distribution networks, in *2011 IEEE International Conference on Communications Workshops (ICC)* (IEEE, Piscataway, 2011), pp. 1–5
537. F.A. Mohamed, H.N. Koivo, Multiobjective optimization using modified game theory for online management of microgrid. Eur. T. Electr. Power **21**(1), 839–854 (2011)
538. A.-H. Mohsenian-Rad, V.W. Wong, J. Jatskevich, R. Schober, A. Leon-Garcia, Autonomous demand-side management based on game-theoretic energy consumption scheduling for the future smart grid. IEEE Trans. Smart Grid **1**(3), 320–331 (2010)
539. P. Oliveira, T. Pinto, H. Morais, Z.A. Vale, I. Praca, MASCEM – an electricity market simulator providing coalition support for virtual power players, in *2009 15th International Conference on Intelligent System Applications to Power Systems* (IEEE, Curitiba, 2009), pp. 1–6
540. T. Pinto, Z.A. Vale, H. Morais, I. Praca, C. Ramos, Multi-agent based electricity market simulator with VPP: Conceptual and implementation issues, in *2009 IEEE Power & Energy Society General Meeting* (IEEE, Calgary, 2009), pp. 1–9

541. I. Praca, C. Ramos, Z. Vale, M. Cordeiro, Mascem: a multiagent system that simulates competitive electricity markets. IEEE Intell. Syst. **18**(6), 54–60 (2003)
542. I. Praça, H. Morais, M. Cardoso, C. Ramos, Z. Vale, Virtual power producers integration into mascem. Establishing Found. Collaborative Netw. **243**, 291–298 (2007)
543. Z. Vale, T. Pinto, I. Praca, H. Morais, MASCEM: electricity markets simulation with strategic agents. IEEE Intell. Syst. **26**(2), 9–17 (2011)
544. Z. Vale, T. Pinto, H. Morais, I. Praca, P. Faria, VPP's multi-level negotiation in smart grids and competitive electricity markets, in *2011 IEEE Power and Energy Society General Meeting* (IEEE, Detroit, 2011), pp. 1–8
545. I. Praca, H. Morais, C. Ramos, Z. Vale, H. Khodr, Multi-agent electricity market simulation with dynamic strategies & virtual power producers, in *2008 IEEE Power and Energy Society General Meeting - Conversion and Delivery of Electrical Energy in the 21st Century* (IEEE, Piscataway, 2008), pp. 1–8
546. M. Andreasson, D.V. Dimarogonas, K.H. Johansson, H. Sandberg, Distributed vs. centralized power systems frequency control, in *2013 European Control Conference (ECC)* (IEEE, Piscataway, 2013), pp. 3524–3529
547. M. Andreasson, D.V. Dimarogonas, H. Sandberg, K.H. Johansson, Distributed PI-control with applications to power systems frequency control, in *2014 American Control Conference (ACC)* (IEEE, Piscataway, 2014), pp. 3183–3188
548. A. Zidan, E.F. El-Saadany, A cooperative multiagent framework for self-healing mechanisms in distribution systems. IEEE Trans. Smart Grid **3**(3), 1525–1539 (2012)
549. J.M. Solanki, N.N. Schulz, Multi-agent system for islanded operation of distribution systems, in *2006 IEEE PES Power Systems Conference and Exposition* (IEEE, Atlanta, 2006), pp. 1735–1740
550. Z. Jiang, Agent-based control framework for distributed energy resources microgrids, in *IEEE/WIC/ACM International Conference on Intelligent Agent Technology, 2006. IAT '06* (IEEE, Hong Kong, 2006), pp. 646–652
551. S. Rivera, A.M. Farid, K. Youcef-Toumi, Chapter 15 – a multi-agent system coordination approach for resilient self-healing operations in multiple microgrids, in *Industrial Agents*, ed. by P.L. Karnouskos (Morgan Kaufmann, Boston 2015), pp. 269–285. http://amfarid.scripts.mit.edu/resources/Books/SPG-BC01.pdf
552. S. Rivera, A.M. Farid, K. Youcef-Toumi, A multi-agent system transient stability platform for resilient self-healing operation of multiple microgrids, in *9th Carnegie Mellon University Electricity Conference* (IEEE, Pittsburgh, 2014), pp. 1–38
553. J. Queiroz, P. Leitão, A. Dias, A multi-agent system for micro grids energy production management supported by predictive data analysis, in *DSIE' 6*, p. 151 (2016)
554. M. Merdan, W. Lepuschitz, B. Groessing, M. Helbok, Process rescheduling and path planning using automation agents, in *Recent Advances in Robotics and Automation* (Springer, Berlin, 2013), pp. 71–80
555. C. Colson, M. Nehrir, R. Gunderson, Distributed multi-agent microgrids: a decentralized approach to resilient power system self-healing, in *2011 4th International Symposium on Resilient Control Systems* (IEEE, Boise, 2011), pp. 83–88
556. N. Cai, N.T.T. Nga, J. Mitra, Economic dispatch in microgrids using multi-agent system, in *2012 North American Power Symposium (NAPS)* (IEEE, Champaign, 2012), pp. 1–5
557. Y. Xu, W. Liu, Novel multiagent based load restoration algorithm for microgrids. IEEE Trans. Smart Grid **2**(1), 152–161 (2011)
558. C.M. Colson, M.H. Nehrir, Algorithms for distributed decision-making for multi-agent microgrid power management, in *2011 IEEE Power and Energy Society General Meeting* (IEEE, Detroit, 2011), pp. 1–8
559. C.M. Colson, M.H. Nehrir, Agent-based power management of microgrids including renewable energy power generation, in *2011 IEEE Power and Energy Society General Meeting* (IEEE, Detroit, 2011), pp. 1–3
560. T. Logenthiran, D. Srinivasan, A.M. Khambadkone, H.N. Aung, Multi-agent system (MAS) for short-term generation scheduling of a microgrid, in *2010 IEEE International Conference*

on Sustainable Energy Technologies (ICSET) (IEEE Computer Society, Singapore, 2010), pp. 1–6. http://dx.doi.org/10.1109/ICSET.2010.5684943

561. J.A. Pecas Lopes, C.L. Moreira, A.G. Madureira, Defining control strategies for analysing microgrids islanded operation, in *2005 IEEE Russia Power Tech* (IEEE, Piscataway, 2005), pp. 1–7
562. J.A. Peas Lopes, C.L. Moreira, A.G. Madureira, Defining control strategies for MicroGrids islanded operation. IEEE Trans. Power Syst. **21**(2), 916–924 (2006)
563. V. Vyatkin, G. Zhabelova, N. Higgins, M. Ulieru, K. Schwarz, N.-K.C. Nair, Standards-enabled smart grid for the future EnergyWeb, in *2010 Innovative Smart Grid Technologies (ISGT)* (Gaithersburg, IEEE Computer Society, 2010), pp. 1–9. http://dx.doi.org/10.1109/ISGT.2010.5434776
564. R. Bi, M. Ding, T.T. Xu, Design of common communication platform of microgrid, in *The 2nd International Symposium on Power Electronics for Distributed Generation Systems* (IEEE, Hefei, 2010), pp. 735–738
565. A.M. Farid, Multi-agent system design principles for resilient operation of future power systems, in *IEEE International Workshop on Intelligent Energy Systems* (IEEE, San Diego, 2014), pp. 1–7. http://engineering.dartmouth.edu/liines/resources/Conferences/SPG-C42.pdf
566. G. Zhabelova, V. Vyatkin, Multiagent Smart Grid Automation Architecture Based on IEC 61850/61499 Intelligent Logical Nodes (2012)
567. N. Higgins, V. Vyatkin, N. Nair, K. Schwarz, Distributed power system automation with IEC 1850, IEC 61499, and intelligent control. IEEE Trans. Syst. Man Cybern. Part C Appl. Rev. **41**(1), 81–92 (2011)
568. J. Lagorse, D. Paire, A. Miraoui, A multi-agent system for energy management of distributed power sources. Renew. Energy **35**(1), 174–182 (2010)
569. T. Logenthiran, D. Srinivasan, Multi-agent system for the operation of an integrated microgrid. J. Renewable Sustainable Energy **4**(1), 013116 (2012)
570. T. Logenthiran, D. Srinivasan, A.M. Khambadkone, H.N. Aung, Multiagent system for real-time operation of a microgrid in real-time digital simulator. IEEE Trans. Smart Grid **3**(2), 925–933 (2012)
571. C.-X. Dou, B. Liu, Multi-agent based hierarchical hybrid control for smart microgrid. IEEE Trans. Smart Grid **4**(2), 771–778 (2013)
572. C.M. Colson, M.H. Nehrir, Comprehensive real-time microgrid power management and control with distributed agents. IEEE Trans. Smart Grid **4**(1), 617–627 (2013)
573. N. Cai, X. Xu, J. Mitra, A hierarchical multi-agent control scheme for a black start-capable microgrid, in *Power and Energy Society General Meeting, 2011 IEEE* (2011), pp. 1–7
574. W. Khamphanchai, S. Pisanupoj, W. Ongsakul, M. Pipattanasomporn, A multi-agent based power system restoration approach in distributed smart grid, in *2011 International Conference and Utility Exhibition on Power and Energy Systems: Issues & Prospects for Asia (ICUE)* (IEEE, New York, 2011), pp. 1–7
575. D. Brandl, Distributed Controls in the Internet of Things Create Control Engineering Resources. Control Engineering, Tech. Rep., 2014. https://www.controleng.com/single-article/distributed-controls-in-the-internet-of-things-create-control-engineering-resources
576. Energy Independence and Security Act (EISA), *Energy Independence and Security Act of 110th United States Congress*, vol. 2007 (2007)
577. C. Greer, D.A. Wollman, D.E. Prochaska, P.A. Boynton, J.A. Mazer, C.T. Nguyen, G.J. FitzPatrick, T.L. Nelson, G.H. Koepke, A.R. Hefner Jr et al., NIST Framework and Roadmap for Smart Grid Interoperability Standards, Release 3.0. NIST, Tech. Rep., 2014
578. R. Ambrosio, S. Widergren, A framework for addressing interoperability issues, in *2007 IEEE Power Engineering Society General Meeting* (IEEE, Piscataway, 2007), pp. 1–5
579. O.C.A.W. Group et al., Openfog architecture overview, in *White Paper OPFWP001*, vol. 216, p. 35 (2016)
580. O.C.A.W. Group et al., Openfog reference architecture for fog computing, in *OPFRA001*, vol. 20817, p. 162, (2017)
581. A. Munir, P. Kansakar, S.U. Khan, IFCIoT: Integrated fog cloud IoT: a novel architectural paradigm for the future internet of things. IEEE Consu. Electron. Mag. **6**(3), 74–82 (2017)

582. AWS, AWS IoT: Developer Guide. Amazon Web Services, Inc., Tech. Rep., 2018
583. SAP, IoT joint reference architecture from INTEL and SAP. SAP/INTEL. Tech. Rep., 2018
584. A. Kumar, How to design the architecture and develop an IoT application on SAP HANA cloud platform (2018). https://sapinsider.wispubs.com/Assets/Articles/2018/June/SPJ-How-to-Design-the-Architecture-and-Develop-an-IoT-Application-on-SAP-HANA-Cloud-Platform
585. CISCO, CISCO IoT Networking. CISCO, Tech. Rep., 2014
586. Microsoft, Microsoft Azure IoT Reference Architecture. Microsoft, Tech. Rep., 2018
587. S. Rohjans, M. Uslar, R. Bleiker, J. González, M. Specht, T. Suding, T. Weidelt, Survey of smart grid standardization studies and recommendations, in *2010 First IEEE Intern. Conference on Smart Grid Communications (SmartGridComm)* (IEEE, Gaithersburg, 2010), pp. 583–588
588. M. Uslar, M. Specht, C.Dänekas, J. Trefke, S. Rohjans, J.M. González, C. Rosinger, R. Bleiker, *Standardization in Smart Grids: Introduction to IT-Related Methodologies, Architectures and Standards* (Springer, Berlin, 2012)
589. EPRI, *The Value of Direct Access to Connected Devices* (Palo Alto, 2017). https://www.epri.com/#/pages/product/3002007825/
590. SMB Smart Grid Strategic Group (SG3), IEC Smart Grid Standardization Roadmap (2010)
591. IEEE Standards Coordinating Committee 21, IEEE Guide for Smart Grid Interoperability of Energy Technology and Information Technology Operation with the Electric Power System (EPS), End-Use Applications, and Loads (2011)
592. V.K.L. Huang, D. Bruckner, C.J. Chen, P. Leitão, G. Monte, T.I. Strasser, K.F. Tsang, Past, present and future trends in industrial electronics standardization, in *IECON 2017 – 43rd Annual Conference of the IEEE Industrial Electronics Society* (IEEE, Beijing, 2017), pp. 6171–6178
593. T. Basso, S. Chakraborty, A. Hoke, M. Coddington, IEEE 1547 standards advancing grid modernization, in *2015 IEEE 42nd Photovoltaic Specialist Conference (PVSC)* (IEEE, Piscataway, 2015), pp. 1–5
594. F. Leccese, An overwiev on IEEE Std 2030, in *2012 11th International Conference on Environment and Electrical Engineering (EEEIC)* (IEEE, Piscataway, 2012), pp. 340–345
595. P. McDaniel, S. McLaughlin, Security and privacy challenges in the smart grid. IEEE Secur. Priv. **7**(3), 75–77 (2009)
596. H.C. Pöhls, V. Angelakis, S. Suppan, K. Fischer, G. Oikonomou, E.Z. Tragos, R.D. Rodriguez, T. Mouroutis, RERUM: Building a reliable IoT upon privacy- and security-enabled smart objects, in *2014 IEEE Wireless Communications and Networking Conference Workshops (WCNCW)* (IEEE, Piscataway, 2014), pp. 122–127
597. X. Wang, J. Zhang, E.M. Schooler, M. Ion, Performance evaluation of attribute-based encryption: toward data privacy in the IoT, in *2014 IEEE International Conference on Communications (ICC)* (IEEE, Piscataway, 2014), pp. 725–730
598. A. Alcaide, E. Palomar, J. Montero-Castillo, A. Ribagorda, Anonymous authentication for privacy-preserving IoT target-driven applications. Comput. Secur. **37**, 111–123 (2013)
599. A. Ukil, S. Bandyopadhyay, A. Pal, Privacy for IoT: Involuntary privacy enablement for smart energy systems, in *2015 IEEE International Conference on Communications (ICC)* (IEEE, Piscataway, 2015), pp. 536–541
600. G. Hug, J.A. Giampapa, Vulnerability assessment of AC state estimation with respect to false data injection cyber-attacks. IEEE Trans. Smart Grid **3**(3), 1362–1370 (2012)
601. A. Teixeira, G. Dán, H. Sandberg, K.H. Johansson, A cyber security study of a scada energy management system: Stealthy deception attacks on the state estimator. IFAC Proc. Vol. **44**(1), 11271–11277 (2011)
602. A. Teixeira, S. Amin, H. Sandberg, K.H. Johansson, S.S. Sastry, Cyber security analysis of state estimators in electric power systems, in *2010 49th IEEE Conference on Decision and Control (CDC)* (IEEE, Piscataway, 2010), pp. 5991–5998
603. D. Minoli, K. Sohraby, B. Occhiogrosso, IoT considerations, requirements, and architectures for smart buildings—energy optimization and next-generation building management systems. IEEE Internet Things J. **4**(1), 269–283 (2017)

604. N. Allen, Cybersecurity weaknesses threaten to make smart cities morecostly and dangerous than their analog predecessors, *USApp–American Politics and Policy Blog* (2016)
605. A. Furfaro, L. Argento, A. Parise, A. Piccolo, Using virtual environments for the assessment of cybersecurity issues in IoT scenarios. Simul. Model. Pract. Theory **73**, 43–54 (2017)
606. V.Y. Pillitteri, T.L. Brewer, Guidelines for Smart Grid Cybersecurity. NIST, Tech. Rep., 2014
607. R. Melton, Pacific Northwest Smart Grid Demonstration Project Technology Performance Report Volume 1: Technology Performance, Pacific Northwest National Laboratory (PNNL), Richland, Tech. Rep., 2015
608. B. Chew, B. Feldman, N. Esch, M. Lynch, Utility Demand Response Market Snapshot. Smart Electric Power Alliance, Tech. Rep., 2017
609. E. Fisher, E. Myers, B. Chew, Der Aggregations in Wholesale Markets. Smart Electric Power Alliance, Tech. Rep., 2017
610. S.E. Widergren, D.J. Hammerstrom, Q. Huang, K. Kalsi, J. Lian, A. Makhmalbaf, T.E. McDermott, D. Sivaraman, Y. Tang, A. Veeramany et al., Transactive systems simulation and valuation platform trial analysis. Pacific Northwest National Laboratory (PNNL), Richland, Tech. Rep., 2017
611. D.J. Hammerstrom, R. Ambrosio, T.A. Carlon, J.G. DeSteese, G.R. Horst, R. Kajfasz, L.L. Kiesling, P. Michie, R.G. Pratt, M. Yao et al., Pacific Northwest Gridwise^{TM} Testbed Demonstration Projects; Part I. Olympic Peninsula Project. Pacific Northwest National Lab.(PNNL), Richland, Tech. Rep., 2008
612. R. Cowart, Recommendations on Non-wires Solutions. Memorandum (2012)
613. Anonymous, National energy grid map index (2014). http://www.geni.org/globalenergy/library/national_energy_grid/index.shtml
614. D. Hammerstrom, D. Johnson, C. Kirkeby, Y. Agalgaonkar, S. Elbert, O. Kuchar, M. Marinovici, R. Melton, K. Subbarao, Z. Taylor et al., Pacific northwest smart grid demonstration project technology performance report volume 1: Technology performance, in *PNWD-4445 volume*, vol. 1 (2015)
615. A. Wright, P. De Filippi, (2015) Decentralized blockchain technology and the rise of lex cryptographia. http://dx.doi.org/10.2139/ssrn.2580664
616. I. Eyal, A.E. Gencer, E.G. Sirer, R. Van Renesse, Bitcoin-ng: a scalable blockchain protocol, in *Proceedings of the 13th USENIX Symposium on Networked Systems Design and Implementation (NSDI '16)* (USENIX, Santa Clara, 2016), pp. 45–59
617. E. Mengelkamp, J. Gärttner, K. Rock, S. Kessler, L. Orsini, C. Weinhardt, Designing microgrid energy markets: a case study: the Brooklyn microgrid. Appl. Energy **210**, 870–880 (2018)
618. Y. Zhou, J. Wu, C. Long, Evaluation of peer-to-peer energy sharing mechanisms based on a multiagent simulation framework. Appl. Energy **222**, 993–1022 (2018)
619. D. Cardwell, Solar Experiment Lets Neighbors Trade Energy Among Themselves. New York Times, Tech. Rep., 2017. https://www.nytimes.com/2017/03/13/business/energy-environment/brooklyn-solar-grid-energy-trading.html
620. U. Papajak, Can The Brooklyn Microgrid Project Revolutionise the Energy Market? Energy Transition: The Global Energiewende, Tech. Rep., 2017. https://energytransition.org/2018/01/can-the-brooklyn-microgrid-project-revolutionise-the-energy-market/
621. P. Ledger, Power ledger white paper, in *Power Ledger Pty Ltd, Australia, White Paper*, 2017
622. T. Sousa, T. Soares, P. Pinson, F. Moret, T. Baroche, E. Sorin, Peer-to-peer and community-based markets: a comprehensive review. Preprint, arXiv:1810.09859 (2018)
623. A. Rutkin, Blockchain aids solar sales. New Sci. **231**(3088), 22 (2016)
624. B. Potter, Blockchain power trading platform to rival batteries (2016)
625. M. Skilton, F. Hovsepian, Impact of security on intelligent systems, in *The 4th Industrial Revolution* (Springer, Berlin, 2018), pp. 207–226
626. V. Venugopalan, C.D. Patterson, Surveying the hardware trojan threat landscape for the internet-of-things. J. Hardw. Syst. Secur. **2**(2), 131–141 (2018)
627. A. Wegner, J. Graham, E. Ribble, A new approach to cyberphysical security in industry 4.0, in *Cybersecurity for Industry 4.0* (Springer, Berlin, 2017), pp. 59–72

628. A. Ouaddah, A.A. Elkalam, A.A. Ouahman, Towards a novel privacy-preserving access control model based on blockchain technology in IoT, in *Europe and MENA Cooperation Advances in Information and Communication Technologies* (Springer, Berlin, 2017), pp. 523–533.
629. M.R. Abdmeziem, F. Charoy, Fault-tolerant and scalable key management protocol for IoT-based collaborative groups, in *Security and Privacy in Communication Networks: SecureComm 2017 International Workshops, ATCS and SePrIoT, Niagara Falls, ON, Canada, October 22–25, 2017, Proceedings 13* (Springer, Berlin, 2018), pp. 320–338
630. N. Baracaldo, L.A.D. Bathen, R.O. Ozugha, R. Engel, S. Tata, H. Ludwig, Securing data provenance in internet of things (IoT) systems, in *International Conference on Service-Oriented Computing* (Springer, Berlin, 2016), pp. 92–98
631. M. Niranjanamurthy, B. Nithya, S. Jagannatha, Analysis of blockchain technology: pros, cons and SWOT. Cluster Comput. 1–15 (2018). http://dx.doi.org/10.1007/s10586-018-2387-5
632. Y. Zhang, J. Wen, The IoT electric business model: using blockchain technology for the internet of things. Peer-to-Peer Netw. Appl. **10**(4), 983–994 (2017)
633. O. Odiete, R.K. Lomotey, R. Deters, Using blockchain to support data and service management in IoV/IoT, in *International Conference on Security with Intelligent Computing and Big-data Services* (Springer, Berlin, 2017), pp. 344–362
634. Q. Xu, K.M.M. Aung, Y. Zhu, K.L. Yong, A blockchain-based storage system for data analytics in the internet of things, in *New Advances in the Internet of Things* (Springer, Berlin, 2018), pp. 119–138
635. Y. Jiang, China's water security: current status, emerging challenges and future prospects. Environ. Sci. Pol. **54**, 106–125 (2015). http://www.sciencedirect.com/science/article/pii/S1462901115300095
636. K.W. Costello, R.C. Hemphill, Electric utilities' 'death spiral': hyperbole or reality? Electr. J. **27**(10), 7–26 (2014)
637. P.R. Newbury, Creative destruction and the natural monopoly'Death Spiral': can electricity distribution utilities survive the incumbent's curse? in *In EURAM13: 13th Annual Conference of the European Academy of Management 2013. European Academy of Management (EURAM).*, 2013
638. N.D. Laws, B.P. Epps, S.O. Peterson, M.S. Laser, G.K. Wanjiru, On the utility death spiral and the impact of utility rate structures on the adoption of residential solar photovoltaics and energy storage. Appl. Energy **185**, 627–641 (2017)
639. J. St. John, As California prepares for wholesale distributed energy aggregation, new players seek approval (2017). https://www.greentechmedia.com/articles/read/california-companies-are-vying-for-aggregated-distributed-energy#gs.IkZGkc8
640. California Independent System Operator (CAISO), Distributed Energy Resource Provider Participation Guide with Checklist. California Independent System Operator (CAISO), Tech. Rep., 2016
641. S.E.D. Coalition, Mapping demand response in Europe today. Tracking Compliance with Article 15, 2014
642. A.M. Prostejovsky, Increased observability in electric distribution grids: emerging applications promoting the active role of distribution system operators. Ph.D. dissertation, Technical University of Denmark, Department of Electrical Engineering, 2017
643. K.A. McDermott, Utility of the future: a vision of a two-way street. Electr. J. **30**(1), 8–16 (2017)
644. J. DEIGN, 15 firms leading the way on energy blockchain. https://www.greentechmedia.com/articles/read/leading-energy-blockchain-firms#gs.SKiyUqc
645. N. Jin, P. Flach, T. Wilcox, R. Sellman, J. Thumim, A. Knobbe, Subgroup discovery in smart electricity meter data. IEEE Trans. Ind. Inf. **10**(2), 1327–1336 (2014)
646. Eurelectric, The Role of DSOS on Smart Grids and Energy Efficiency. Eurelectric, Tech. Rep., 2012

647. EFET, The Role and Responsibilities of Electricity Distribution System Operators (DSOS), Particularly Regarding Access to Storage. The European Federation of Energy Traders, Tech. Rep., 2014
648. A. Monti, F. Ponci, M. Ferdowsi, P. McKeever, A. Löwen, Towards a new approach for electrical grid management: the role of the cloud, in *2015 IEEE International Workshop on Measurements & Networking (M&N)* (IEEE, Piscataway, 2015), pp. 1–6
649. P.D. Lund, J. Lindgren, J. Mikkola, J. Salpakari, Review of energy system flexibility measures to enable high levels of variable renewable electricity. Renew. Sust. Energ. Rev. **45**, 785–807 (2015)
650. A.T. de Almeida, F.J. Ferreira, D. Both, Technical and economical considerations in the application of variable speed drives with electric motor systems, in *2004 IEEE Industrial and Commercial Power Systems Technical Conference* (IEEE, Piscataway, 2004), pp. 136–144
651. M. Åhman, L.J. Nilsson, Decarbonizing industry in the EU: climate, trade and industrial policy strategies, in *Decarbonization in the European Union* (Springer, Berlin, 2015), pp. 92–114.
652. M. Åhman, L.J. Nilsson, B. Johansson, Global climate policy and deep decarbonization of energy-intensive industries. Clim. Policy **17**(5), 634–649 (2017)
653. M. Kramer, The link between competitive advantage and corporate social responsibility. Harv. Bus. Rev. 6–20 (2006)
654. A.B. Carroll, Corporate social responsibility: the centerpiece of competing and complementary frameworks. Organ. Dyn. **44**(2), 87–96 (2015)
655. W.C. Schoonenberg, A.M. Farid, A Dynamic model for the energy management of microgrid-enabled production systems. J. Clean. Prod. **1**(1), 1–10 (2017)
656. B. Jiang, A.M. Farid, K. Youcef-Toumi, A comparison of day-ahead wholesale market: social welfare vs industrial demand side management, in *IEEE International Conference on Industrial Technology* (IEEE, Sevilla, 2015), pp. 1–7. http://dx.doi.org/10.1109/ICIT.2015.7125502
657. B. Jiang, A. Muzhikyan, A.M. Farid, K. Youcef-Toumi, Impacts of industrial baseline errors in demand side management enabled enterprise control, in *IECON 2015 – 41st Annual Conference of the IEEE Industrial Electronics Society* (IEEE, Yokohama, 2015), pp. 1–6. http://dx.doi.org/10.1109/IECON.2015.7392637
658. B. Leukert, T. Kubach, C. Eckert, IoT 2020: Smart and Secure IoT Platform. International Electrotechnical Commission, Tech. Rep., 2016
659. EIA, Electricity monthly update (2018). https://www.eia.gov/electricity/monthly/update/archive/march2018/end_use.php
660. Z. Ma, J.D. Billanes, B.N. Jørgensen, Aggregation potentials for buildings—business models of demand response and virtual power plants. Energies **10**(10), 1646 (2017)
661. Google, Achieving Our 100% Renewable Energy Purchasing Goal and Going Beyond. Google, Tech. Rep., 2016
662. Google, Environmental Report. Google, Tech. Rep., 2016
663. Walmart, 2018 Global Responsibility Report Summary. Walmart, Tech. Rep., 2018
664. O. Vermesan, P. Friess, *Internet of Things: Converging Technologies for Smart Environments and Integrated Ecosystems* (River Publishers, Aalborg, 2013)
665. Solar Energy Industries Associations, Net metering (2017). https://www.seia.org/initiatives/net-metering
666. M. Maasoumy, A. Sangiovanni-Vincentelli et al., Smart connected buildings design automation: foundations and trends. Found. Trends® Electr. Design Automa. **10**(1–2), 1–143 (2016)
667. E. Stuart, P.H. Larsen, J.P. Carvallo, C.A. Goldman, D. Gilligan, US Energy Service Company (ESCO) Industry: Recent Market Trends. Lawrence Berkeley National Lab. (LBNL), Berkeley, Tech. Rep., 2016
668. V. Marinakis, H. Doukas, An advanced IoT-based system for intelligent energy management in buildings. Sensors **18**(2), 610 (2018)
669. A. Herceg, Beyond the Walls: Benchmarking BEMS Software and Hardware. Luxresearch, Tech. Rep., 2015

670. U.S. Department of Energy, The internet of things (IoT) and Energy Management in the Modern Building. U.S. Department of Energy, Tech. Rep., 2016
671. Luxresearch, Sensors and Controls for BEMS: Providing the Neural Network to Net-Zero Energy. Luxresearch, Tech. Rep., 2012
672. Incenergy, Security in Internet-Connected Building Automation and Energy Management Systems. Incenergy, Tech. Rep., 2014
673. S. Aman, Y. Simmhan, V.K. Prasanna, Energy management systems: state of the art and emerging trends. IEEE Commun. Mag. **51**(1), 114–119 (2013)

Index

S. O. Muhanji et al., *eIoT*, https://doi.org/10.1007/978-3-030-10427-6

www.ingramcontent.com/pod-product-compliance
Lightning Source LLC
LaVergne TN
LVHW082234200926
848484LV00008B/2356